Essential Knowledge for Transistor-Level LSI Circuit Design

Toru Nakura

Essential Knowledge for Transistor-Level LSI Circuit Design

Springer

Toru Nakura
The University of Tokyo
Tokyo, Japan

Translated by Yuta Toriyama

Original Japanese language edition published by University of Tokyo Press.
LSI Sekkei Joshiki Koza

ISBN 978-981-10-0423-0 ISBN 978-981-10-0424-7 (eBook)
DOI 10.1007/978-981-10-0424-7

Library of Congress Control Number: 2016932877

Printed on acid-free paper

This Springer imprint is published by Springer Nature
The registered company is Springer Science+Business Media Singapore Pte Ltd.

Preface

You have learned about the transistor in the classroom. You also know the operating principles of an op amp. You might have even read Razavi's textbook and mastered the basics of analog circuits. However, you might still be at a loss when you are told, "OK, now design an op amp."

To design an LSI that actually functions and to be able to properly measure it, an extremely large amount of miscellaneous, detailed knowledge is necessary. This knowledge is transferred from seniors to juniors by word of mouth within university laboratories and development groups in companies. The handing down of technology from seniors to juniors through repeated failures is a beautiful tradition, but sadly the times do not allow engineers to be at such leisure. This text is a collection of the miscellaneous knowledge essential for circuit designers, which has not been in textbooks before, summarized as the issues that need to be considered in each design step.

The first half of this text explains important design issues such as the operating principles of CAD tools. CAD tools are used to conduct LSI circuit design, and it is necessary to accurately reflect the above-mentioned "issues that need to be considered in each design step" within the CAD tool as data. Here, there is actually a mixture of "issues small enough to be ignored as inconsequential" and "issues that may seem at first negligible but actually are very important." To make accurate distinctions between these cases and to design a properly operating LSI, it is necessary to precisely understand what each of the settings in a CAD tool means, while keeping in consideration the operating principles of the CAD tool itself. When I was in my fourth year of undergraduate study and was first assigned to a laboratory, my supervisor, Professor Asada, told me: "Never trust simulators other than the ones you have made yourself."

This means that the calculation results of simulators should always be taken with a grain of salt, since various errors arise depending on the modeling methodology, unreasonable assumptions can be built in, and users can give boundary conditions which the author of the tool had not considered. Times have changed, and now not only is it impossible to design without using commercial CAD tools, but also there is a growing sense that "the ability to design circuits = the ability to use CAD tools."

Even so, while it may not be possible to design a simulator from scratch, I believe that the minimum requirement for circuit designers is to understand the principles of the simulators that are being used. For example, almost all circuit designers have simulated a ring oscillator with an odd number of inverters and wondered why the circuit nodes are all at VDD/2 instead of oscillating. Here, if the designer does not understand how the simulator works, he may draw the incorrect conclusion that “an odd number of inverters in a ring does not oscillate.” That is, it is important not to blindly believe the simulator results even though the problem solved was not well suited for the simulator. In addition, it is necessary to be confident with and have a good command of the various types of CAD tools, and to be able to make accurate decisions from the implications of the calculation results.

In the latter half of this text, necessary knowledge for measuring your own LSI chips is explained. For example, there are two types of oscilloscopes, sampling and real-time, and accurate measurements cannot be made without understanding the operating principles and features of each.

The operating principles of a transistor, the gains of common source and common drain circuits, the circuit techniques of operational amplifiers, A/D converters and PLLs, the details of CAD use, and so on and so forth should be studied with their respective textbooks and documents. This textbook serves as a complement to these ideas, and should be read to acquire knowledge that will become the platform for circuit design. It is assumed that the reader knows some basics, such as transistor characteristics.

The aim of this book is to be useful for newcomers to a lab, or fresh graduates who are assigned to a circuit design group but have little experience in circuit design. This book is also ideal for those who have some experience in circuit design to confirm and complement the knowledge that they already have.

Tokyo, Japan Toru Nakura

Contents

1 Schematic Entry 1
1.1 Schematic Entry 1
1.1.1 The Body Terminal and the Well Structure 1
1.1.2 Transistor Parameters 3
1.2 Models and Parameters 6
1.2.1 Physical Phenomena, the Model, and Parameters 6
1.2.2 Model Equations 7
1.2.3 SPICE Parameters 12
1.2.4 Understanding the Model 14
1.3 Techniques for Circuit Design 15
1.3.1 Hierarchical Design 15
1.3.2 Dealing with Supply and Ground 16
1.3.3 Connections by Labeling 17
1.3.4 Connection Points 18

2 SPICE Simulation 19
2.1 Principles of Simulation 19
2.1.1 DC Analysis 19
2.1.2 Linear Circuit Elements 19
2.1.3 Nonlinear Circuit Elements 21
2.1.4 AC Analysis 23
2.1.5 Transient Analysis 25
2.1.6 Harmonic Balance Analysis 30
2.1.7 Analysis Method Characteristics and Comparison 31
2.2 Fast SPICE 32
2.2.1 Partitioning and Event-Driven Simulation 33
2.2.2 Unpartitionable Circuits 35
2.2.3 Time Step Control 35
2.2.4 Simplification of the Model 36
2.2.5 Automatic Decision and Specification of Simulation Accuracy 36

2.3 A Simple HSPICE Manual 37
2.3.1 Basic Points 37
2.3.2 Defining Elements 38
2.3.3 Voltage and Current Sources 40
2.3.4 Simulation Types 42
2.3.5 File Includes and Libraries 43
2.3.6 Options and the `.MEASURE` Command 43

3 Layout and Verification 49
3.1 The Basic Process of LSI Fabrication 49
3.1.1 The Three-Dimensional LSI Structure 49
3.1.2 Photolithography 50
3.1.3 Deposition 53
3.1.4 Removal of Unnecessary Parts 53
3.1.5 Introduction of Impurities 55
3.1.6 CMOS Fabrication Process 55
3.1.7 Dual Damascene 55
3.2 Design Rules 58
3.2.1 Basic Rules 58
3.2.2 The Grid 59
3.2.3 Density Rules 59
3.2.4 Dummy Transistors 61
3.2.5 Antenna Rules 62
3.2.6 Electromigration 62
3.2.7 Hand-Drawn Layers and Autogenerated Layers 63
3.3 Basic Layout 64
3.3.1 Layout of Transistors 64
3.3.2 Layout of Resistors 64
3.3.3 Layout of Capacitors 66
3.3.4 Layout of Inductors 67
3.4 Layout Editors 67
3.4.1 Layers 68
3.4.2 Display and Grid 68
3.4.3 Objects 69
3.5 Layout Know-How 70
3.5.1 Layout Editor Settings 70
3.5.2 Hierarchical Layout 71
3.5.3 Double Back 71
3.5.4 Supply Lines 72
3.5.5 Clock Distribution 73
3.5.6 Shields 74
3.6 Layout Verification 75
3.6.1 DRC 75
3.6.2 LVS 76
3.6.3 ERC 79

3.6.4 Antenna Check ... 80
3.6.5 Density Check ... 80
3.6.6 Verification Types and Order ... 80
3.6.7 Flat Verification and Hierarchical Verification ... 81

4 Interconnect RC Extraction ... 83
4.1 Parasitic Resistance and Parasitic Capacitance ... 83
4.2 Principles of RC Extraction Tools ... 84
4.2.1 Resistance Extraction ... 84
4.2.2 Resistance Measurement ... 86
4.2.3 Capacitance Extraction ... 86
4.3 AD/AS/PD/PS and HDIF ... 91
4.4 Typical Options ... 93
4.4.1 C Extraction and RC Extraction ... 93
4.4.2 Compaction ... 94
4.4.3 Dealing with Cross-Coupled Capacitances ... 95
4.4.4 Dealing with Supply Lines ... 96
4.4.5 Node Specification and Cell Specification ... 97
4.4.6 Dealing with Floating Nodes and Dummies ... 98
4.4.7 XREF with LVS ... 99
4.5 Reduction of Interconnect RC ... 99
4.5.1 Process Technology ... 100
4.5.2 Design Techniques ... 101

5 IO Buffers ... 103
5.1 Signal Path Between Chips ... 103
5.1.1 Pads ... 103
5.1.2 Packages and Bonding Wires ... 104
5.1.3 Transmission Lines ... 106
5.1.4 Termination Methods ... 110
5.1.5 Voltage Levels ... 111
5.2 ESD ... 112
5.2.1 ESD Models ... 112
5.2.2 ESD Protection Circuits ... 113
5.2.3 Miscellaneous Topics Regarding ESD Protection Circuitry ... 114
5.3 Types of IO Buffers and Their Layout ... 115
5.3.1 IO Buffer Examples ... 116
5.3.2 Supply Rings ... 118
5.4 Determining Pin Placement ... 118
5.4.1 Supply Pin ... 119
5.4.2 Shielding ... 119
5.4.3 Symmetry ... 120
5.4.4 Assembly and Measurement ... 120

6 Noise 121
6.1 Types and Causes of Malfunction 121
6.1.1 No Response 121
6.1.2 Timing Errors 122
6.1.3 Analog Errors 124
6.2 Types of Noise and Their Countermeasures 124
6.2.1 PVT Variations 124
6.2.2 Supply Noise 127
6.2.3 Substrate Noise 130
6.2.4 Cross-Talk Noise 132
6.2.5 EMC 133

7 Problems Due To the Progress of Miniaturization 135
7.1 Variation 135
7.1.1 About Variation 135
7.1.2 Types and Causes of Variation 135
7.1.3 The Effects of Variation 137
7.1.4 Monte Carlo Simulations 141
7.2 Leakage Currents 144
7.2.1 Gate Leakage and High-K 145
7.2.2 Subthreshold Leakage 145
7.2.3 Junction Leakage 146
7.3 Degradation of Characteristics 147
7.3.1 Electromigration 147
7.3.2 Stress Migration 148
7.3.3 Soft Errors 148
7.3.4 Hot Carrier Injection 149
7.3.5 NBTI 150
7.3.6 Random Telegraph Noise 151
7.3.7 Simulation of Degradation Prediction 151

8 Measurement Devices 155
8.1 Sources of Signals to the Chip 155
8.1.1 Power Supply 155
8.1.2 Signal Generators 156
8.1.3 Pulse Pattern Generators 158
8.2 Observers of Signals from the Chip 159
8.2.1 Sampling Oscilloscopes 159
8.2.2 Real-Time Oscilloscopes 165
8.2.3 Spectrum Analyzers 167
8.3 Equipment with Both Signal Input and Output 169
8.3.1 BERT 169
8.3.2 Network Analyzers 171
8.3.3 Logic Analyzers 175

9 Measurement Techniques 177
9.1 Supply · Ground and the Return Path 177
9.1.1 Ground 177
9.1.2 Supply and Decoupling Capacitance 178
9.1.3 Return Path 179
9.2 Various Components 181
9.2.1 Connectors and Cables 181
9.2.2 Accessories 183
9.2.3 Probes 186
9.2.4 Assembling Components 187
9.2.5 Shield Room 188
9.3 Assembly Examples 189
9.4 GPIB, Measurement Automation, and C Programming 191

10 The Overall Design Procedure 195
10.1 Before Starting Your Design 195
10.1.1 What Are You Making and Why 195
10.1.2 Determining the Final Image 196
10.1.3 Determining CAD Tools 197
10.2 Checking Transistor Characteristics 198
10.2.1 SPICE Parameters 198
10.2.2 DC Characteristics and Inverter Delay 199
10.3 Checking the General Flow 201
10.3.1 Schematic Editor 201
10.3.2 Inverter Layout and LVS/DRC 202
10.3.3 RC Extraction 203
10.4 Finally, Some Real Design 204
10.4.1 Circuit Design and Considering the Measurement Methodology 204
10.4.2 Layout Design 205
10.5 After Submission of Design Data 207
10.5.1 Preparation for Measurement 207
10.5.2 Preparing Patent Documents 208
10.6 Measurements and Onward 208
10.6.1 Measurements 208
10.6.2 Writing a Report 208
10.6.3 Toward Your Next Design 209

Epilogue 211

Chapter 1
Schematic Entry

Let us first start a CAD tool and draw a circuit diagram. Even those of you who might say "I have never used Linux before," you should create an account, set up your `.cshrc` file, and launch the schematic editor. The moment you place down a transistor symbol, you should wonder: "What is this fourth terminal? The transistors I have seen before in textbooks only had three ..."

1.1 Schematic Entry

1.1.1 The Body Terminal and the Well Structure

The first circuit you design will probably be an inverter or a NAND gate. When a schematic editor is started and the PMOS and NMOS symbols are instantiated, the MOS transistors will be a four-terminal device, as depicted in Fig. 1.1a. This fourth terminal is in fact the body (B) terminal. As shown in Fig. 1.1b, usually this terminal is connected to V_{DD} for a PMOS and G_{ND} for an NMOS. Textbooks often omit this terminal because this connection is obvious. However, because a separate voltage can be applied to the body terminal in certain cases, this connection must be explicitly specified during circuit design.

In order to understand the body terminal, it is necessary to understand the cross-sectional transistor structure, particularly the well structure. This cross-sectional structure is shown in Fig. 1.2. The silicon wafer is usually of a p-type substrate. That shown in Fig. 1.2a is called a double-well structure and is a basic structure for MOS fabrication. An NMOS consists of its source/drain in an n-type region within a P-well, whereas a PMOS is made from its source/drain in a p-type region within an N-well. Usually the entire P-well is biased to G_{ND} and the N-well is biased to V_{DD}. Here, because the p-substrate of the entire chip and the P-well are conductive, body potentials of NMOS devices are common among all transistors. On the other

T. Nakura, *Essential Knowledge for Transistor-Level LSI Circuit Design*,
DOI 10.1007/978-981-10-0424-7_1

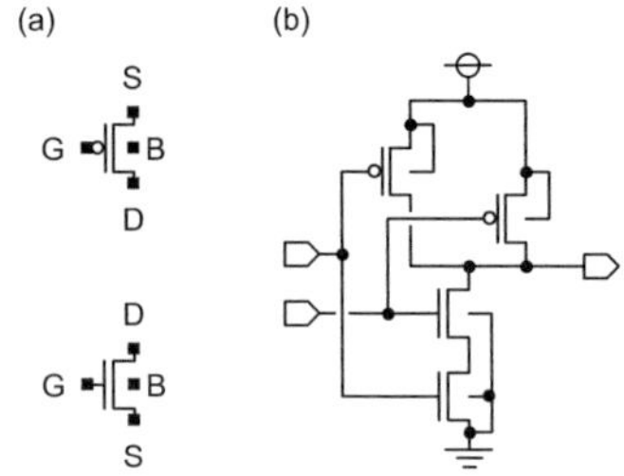

Fig. 1.1 MOS transistor symbol with four terminals

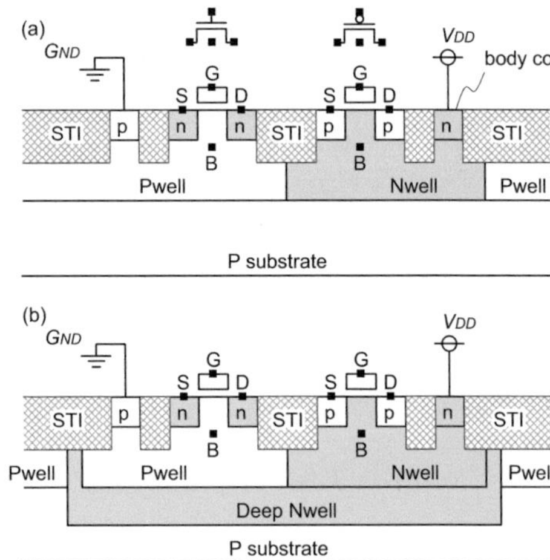

Fig. 1.2 Cross-sectional structure. (**a**) Double-well structure, (**b**) triple-well, deep N-well structure

hand, because the N-well is separated from both the P-well and p-substrate, the body potential of a PMOS device can either be V_{DD} as usual or some other bias potential.

A triple-well or deep N-well (DNW) structure is depicted in Fig. 1.2b. The deep N-well is laid on top of the p-type substrate, and the P-well and N-well where the transistors are constructed are laid on top of the deep N-well. Through this method, the P-well is separated, allowing the voltage to be controlled. Furthermore, because the circuit within the deep N-well is separated from the p-substrate in this structure, there is the benefit that this circuitry is less susceptible to noise that propagates through the p-substrate. This technique of surrounding the entirety of the analog circuit, which is sensitive to noise, with DNW is often used to protect the analog circuitry from noise that digital circuits generate. The specifics of noise will be explained in more detail in Chap. 6. Also, in the figure, STI stands for shallow trench isolation, which is an oxide film used to separate individual transistors.

In any case, the transistor is a four-terminal device, and the body terminal indicates the potential of the well. It is important to understand that usually this terminal is connected to V_{DD} for a PMOS and G_{ND} for an NMOS, but there are cases where the body potential is controlled. For example, if you want to output a voltage of $V_{DD}/2$, it is necessary to bias the body voltage of the PMOS as in Fig. 1.3a.

Fig. 1.3 An example of body voltage biasing

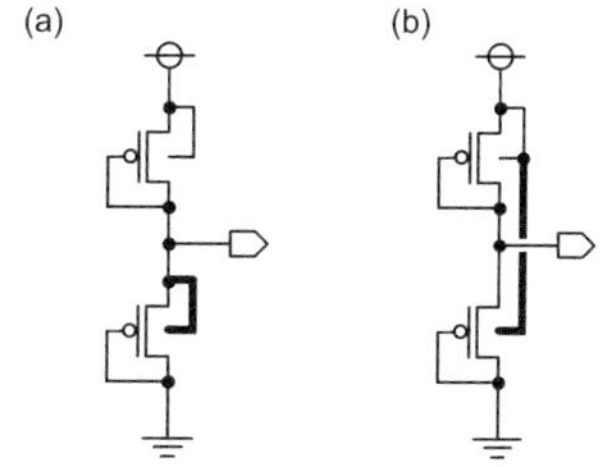

In Fig. 1.3b, the output does not become $V_{DD}/2$, because the characteristics of the PMOS devices on top and bottom differ.

1.1.2 Transistor Parameters

When a transistor is placed down in a schematic editor, it is necessary to specify its characteristics.

1.1.2.1 Gate Length *L*

First the gate length L must be determined. In general, the shortest length allowed in the fabrication process is used. For example, in a $0.18\,\mu$m process, `0.18 um` would be chosen. Some processes specify `0.2 um` in the schematic for ease of layout, while the simulator subtracts $0.02\,\mu$m for simulations. Look for a basic example around you and confirm the above before starting your design.

Also, in analog circuits, a gate length value larger than the minimum possible is sometimes used to mitigate variation or improve the circuit characteristics. The feel for how this goes about will be accumulated through experience in a variety of situations.

1.1.2.2 Gate Width *W* and Multiplier *M*

The next step is to determine the gate width W. A smaller W is desirable for circuit area and power consumption considerations, whereas a larger W yields higher operation speed and better tolerance against variation. You might think that "a wider W will make the load capacitance larger, so the operating speed should be the same." However, for the same wiring length, the relative effect of the parasitic wiring capacitance becomes smaller as shown in Fig. 1.4, thus resulting in a faster operating speed.

As a rule of thumb, the NMOS should have a width W 5$\sim$10 times larger than L, and the PMOS should have a W 2$\sim$2.5 times larger than the NMOS W.

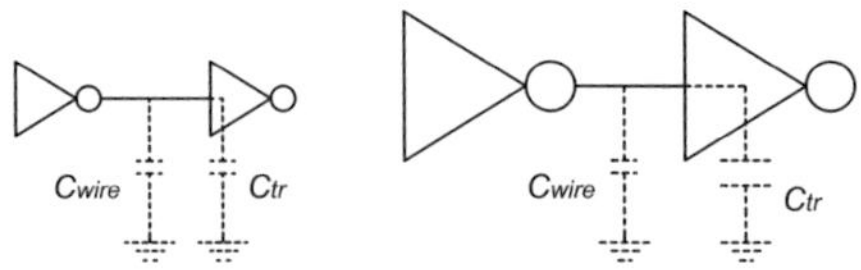

Fig. 1.4 Wider *W* improves operating speed

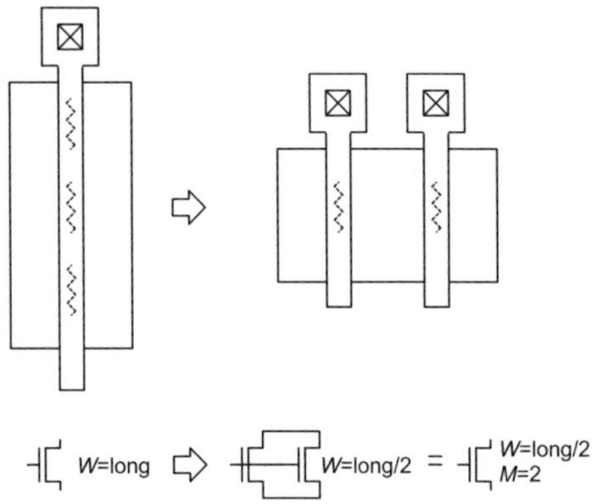

Fig. 1.5 Parallel connection of transistors

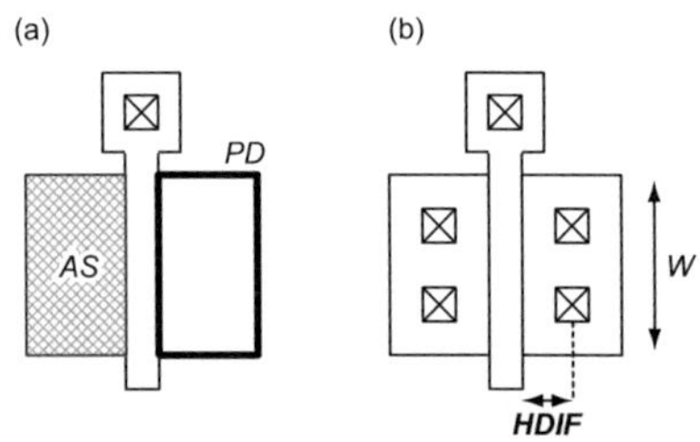

Fig. 1.6 AD, AS, PD, PS, and **HDIF**

As you proceed with the circuit design, you will encounter situations where you would like to use very large *W* transistors. However, in general, there is an upper limit on the maximum usable *W*, and the transistor characteristic is deteriorated due to the increasing series resistance from one end of the gate to the other. Therefore, to obtain a transistor with a large *W*, many short *W* transistors are connected in parallel (Fig. 1.5). In the schematic, rather than placing two transistors in parallel, a single transistor is placed whose parameter *M* (multiplier) is set to be 2. By using the parameter *M*, not only is it easy to change the number of parallel transistors, but also the simulation time is reduced.

1.1.2.3 AD/AS/PD/PS

AD/AS/PD/PS indicates (A) area, (P) periphery, (D) drain, and (S) source. That is, AD is the area of the drain, PS is the peripheral length of the source, and so on (Fig. 1.6). These values are used to calculate capacitances. As indicated in Fig. 1.7,

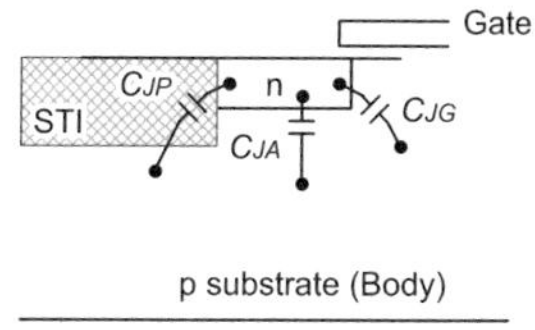

Fig. 1.7 PN junction capacitance

the source and drain regions have capacitance due to a reverse-biased PN junction, which affect the speed and power consumption of the circuitry. The total drain capacitance C_D is expressed as

$$C_D = C_{JA} \times AD + C_{JP} \times (PD - W) + C_{JG} \times W, \tag{1.1}$$

where C_{JA}[F/m^2] is the unit area capacitance of the drain region at the bottom plane, C_{JP}[F/m] is the unit length capacitance of the drain side edges, and C_{JG}[F/m] is the unit length capacitance of the drain gate edge.

Because the area and perimeter of the source and drain regions cannot be deduced from the transistor connectivity, the values for these parameters must be specified explicitly. Otherwise, the simulator may assume these values to be zero, calculate capacitances which are smaller than the actual values, and output optimistic results. However, some transistor models, such as BSIM3, have a specific parameter **HDIF**, which indicates the distance between the gate edge and the source/drain contacts as shown in Fig. 1.6b. This allows the simulator to automatically calculate the area and perimeter values as $A = W \times$ 2**HDIF** and $P =$ 4**HDIF** $+ 2W$, which can be used to calculate Eq. (1.1) without specifying AD/AS/PD/PS. This should be confirmed before starting design, and if the model does not have the **HDIF** parameter, it is necessary to take one of the following actions appropriately:

- Continue design, keeping in mind that the simulated delay is smaller than the actual delay.
- Use a schematic editor that can automatically calculate AD/AS/PD/PS.
- Create your own script to add AD/AS/PD/PS to the netlist generated from the schematic editor.
- Enter the values for AD/AS/PD/PS to each transistor on the schematic.

1.1.2.4 Model

In addition, it is necessary to specify the model for each transistor in the schematic. This will be described in further detail in the next section.

1.2 Models and Parameters

1.2.1 Physical Phenomena, the Model, and Parameters

What does a "model" mean, anyways? When dealing with voltages and currents, all physical phenomena can be analyzed by solving Maxwell's equations under boundary conditions. However, this method requires immense computation time and gives very little insight. Therefore, the physical phenomena are approximated by simple equations. These "simple equations" are the "model."

The very first model you learned to analyze the physical phenomena internal to the transistor is

$$I_D = \frac{1}{2}\frac{W}{L}\mu C_{ox}\left\{2(V_G - V_{th})V_D - V_D^2\right\} \qquad (V_D < V_{th}) \tag{1.2}$$

$$= \frac{1}{2}\frac{W}{L}\mu C_{ox}(V_G - V_{th})^2 \qquad (V_{th} < V_D), \tag{1.3}$$

and these equations describe the relationship between voltage and current as shown in Fig. 1.8a. For a slightly more complicated but more realistic model,

$$I_D = \frac{1}{2}\frac{W}{L}\mu C_{ox}\left\{2(V_G - V_{th})V_D - V_D^2\right\} \qquad (V_D < V_{th}) \tag{1.4}$$

$$= \frac{1}{2}\frac{W}{L}\mu C_{ox}\left\{(V_G - V_{th})^2\right\}(1 + \lambda V_D) \qquad (V_{th} < V_D) \tag{1.5}$$

as shown in Fig. 1.8b, or for a simpler model,

$$I_D = \frac{1}{2}\frac{W}{L}\mu C_{ox}\frac{(V_G - V_{th})^2}{V_{th}}V_D \qquad (V_D < V_{th}) \tag{1.6}$$

$$= \frac{1}{2}\frac{W}{L}\mu C_{ox}(V_G - V_{th})^2 \qquad (V_{th} < V_D) \tag{1.7}$$

as shown in Fig. 1.8c is sometimes enough.

Each analytical expression of Fig. 1.8a–c can be called a "model" which represents the relationship between the currents and voltages of the transistor. Here, μ, C_{ox}, and V_{th} are the "parameters," and I_D and V_D are the actual values of current and voltage and are called "variables."

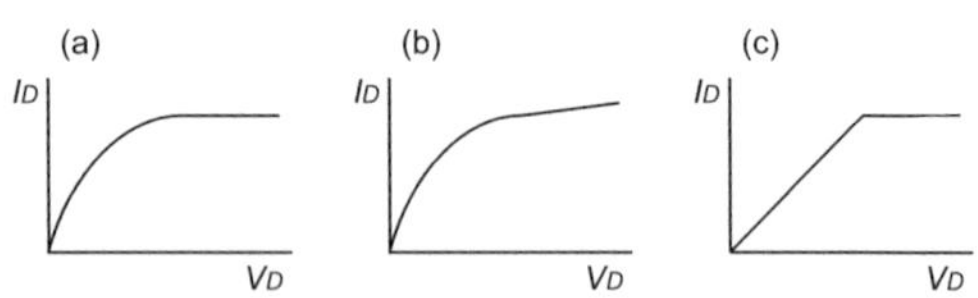

Fig. 1.8 Simple models

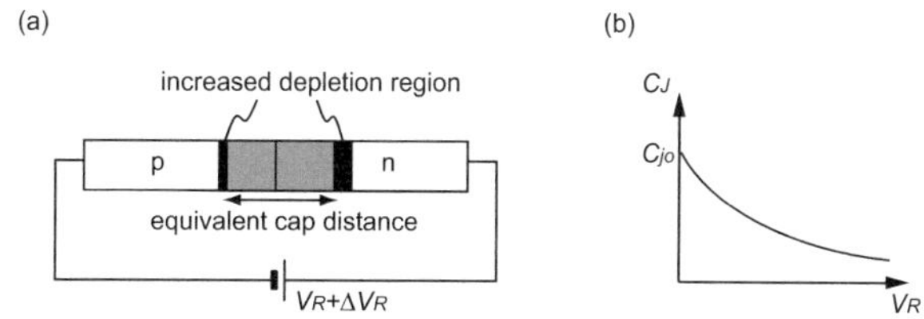

Fig. 1.9 PN junction capacitance

1.2.2 Model Equations

1.2.2.1 PN Junction Capacitance

The source and drain of a transistor have a reverse-biased PN junction capacitance with the well (body). The depletion region width of a PN junction changes with the reverse bias voltage, thus also changing the capacitance value. As indicated in Fig. 1.9, when some applied voltage generates some depletion region width, applying an additional reverse bias voltage will cause charge to accumulate on either end and increase the depletion region. This can be viewed as an equivalent capacitance with the plate separation distance as the original depletion region width. Therefore, when the reverse bias is small, the capacitance is large, and vice versa. For a reverse bias voltage V_R, the differential capacitance is

$$C_J = C_{jo}\left(1 + \frac{V_R}{V_{bi}}\right)^{-m}, \tag{1.8}$$

and the total charge that accumulates when a reverse bias voltage of 0 to V_{DD} is applied to the PN junction is

$$Q = \int_0^{V_{DD}} C_{jo}\left(1 + \frac{V_R}{V_{bi}}\right)^{-m} dV_R. \tag{1.9}$$

Here, V_{bi} is the built-in potential, which is the potential difference across the PN junction when the bias voltage is zero, and C_{jo} is the depletion region capacitance when the bias voltage is zero. The factor m is 1/2 when the doping concentration of the PN junction sees an abrupt change, whereas it is 1/3 when the change is linear. In reality, the value varies between 1/3 and 1/2. Roughly, the capacitance varies as shown in Fig. 1.9b. That is, in Eq. (1.1),

$$C_{JA} = \mathbf{CJ}\left(1 + \frac{V_R}{\mathbf{PB}}\right)^{-\mathbf{MJ}} \tag{1.10}$$

$$C_{JP} = \mathbf{CJSW}\left(1 + \frac{V_R}{\mathbf{PBSW}}\right)^{-\mathbf{MJSW}} \tag{1.11}$$

$$C_{JG} = \mathbf{CJGATE}\left(1 + \frac{V_R}{\mathbf{PHP}}\right)^{-\mathbf{MJGATE}} \tag{1.12}$$

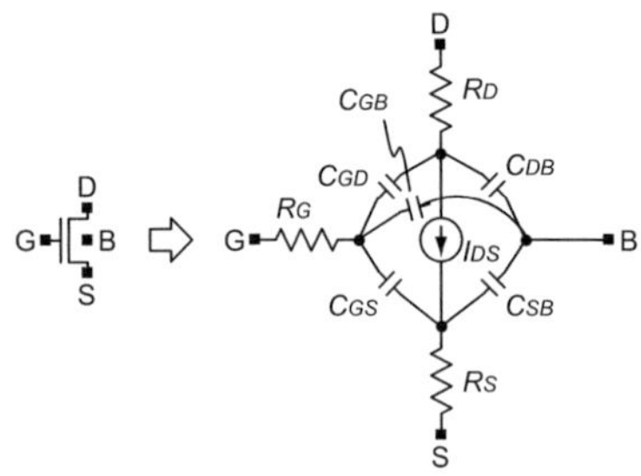

Fig. 1.10 A transistor, modeled as capacitances and a voltage-controlled current source

and the parameters in bold are dictated by the fabrication process. Sometimes, C_{JP} and C_{JG} are treated as a single variable C_{JP}.

1.2.2.2 Drain Current (Model1 ~ BSIM4)

As shown in Fig. 1.10, a transistor is modeled as capacitors and a current source, whose values change as the terminal voltages vary. In other words, the capacitance and current can be expressed as $C(V_G, V_D, V_S, V_B)$ and $I_D(V_G, V_D, V_S, V_B)$. There are a variety of these analytic expressions, or "models." As the number of unignorable physical phenomena increases with the miniaturization of transistors, models including these effects are proposed, and even now, new models are introduced and updated.

The most basic model for the current source is a model called LEVEL1:

$$I_D = 0 \qquad (V_G < V_{th}) \tag{1.13}$$

$$I_D = \frac{W}{L}\mathbf{KP}\left\{(V_G - V_{th})V_D - V_D^2\right\} \qquad (V_D < V_G - V_{th}) \tag{1.14}$$

$$I_D = \frac{1}{2}\frac{W}{L}\mathbf{KP}\left\{\{(V_G - V_{th})^2\right\}(1 + \mathbf{LAMBDA}\,V_D)\ (V_G - V_{th} < V_D) \tag{1.15}$$

$$V_{th} = V_{bi} + \mathbf{GAMMA}\sqrt{(\mathbf{PHI} + V_{SB})} \tag{1.16}$$

$$V_{bi} = \mathbf{VTH0} - \mathbf{GAMMA}\sqrt{\mathbf{PHI}} \tag{1.17}$$

as is expressed in Eqs. (1.4) and (1.5).

When the transistor was first developed, a model of this level of accuracy was enough. Since then, more complex models such as LEVEL2, which includes threshold voltage variation, mobility degradation due to the vertical electric field, and subthreshold characteristics, and LEVEL3, which includes short-channel effects such as drain-induced barrier lowering (DIBL), have been developed. Also, the BSIM model, considered to be the second-generation model, and its expanded versions BSIM2, BSIM3, and BSIM4 models are widely in use.

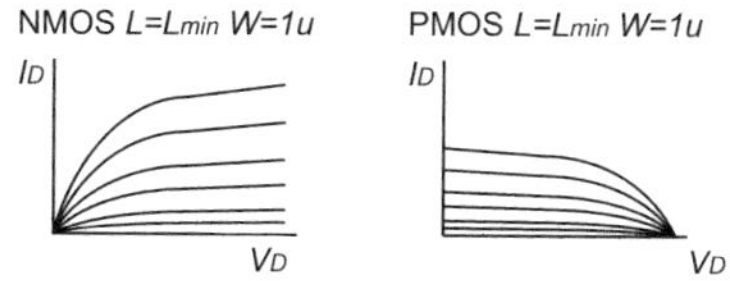

Fig. 1.11 Transistor DC characteristics

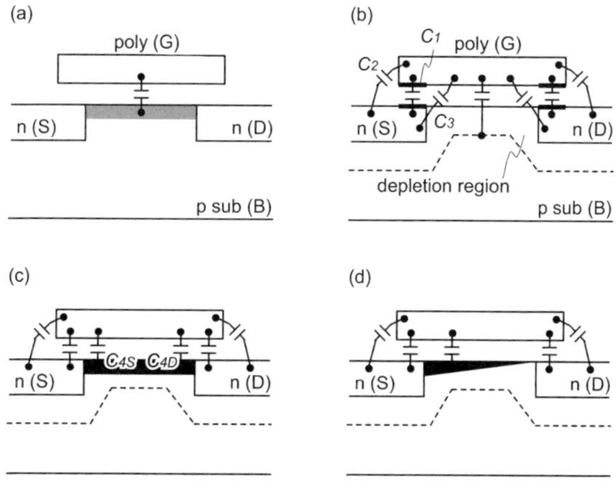

Fig. 1.12 Gate capacitance models in various operating regions of the transistor. (**a**) Accumulation. (**b**) Off state. (**c**) Linear region. (**d**) Saturation region

LEVEL1 and LEVEL2 models are comprehensible to the regular circuit designer, but it is impossible to understand everything that is modeled in BSIM3 or BSIM4 because these models are too complicated. Therefore, an intuitive circuit design should be conducted with basic I_D characteristics based on simulations as shown in Fig. 1.11, after which the values of L and W are iteratively optimized until convergence in simulation.

1.2.2.3 Transistor Capacitance

The capacitance model of the transistor is equally important to the current source model. The drain-body and source-body PN junction capacitances can be expressed by Eqs. (1.1), (1.10), (1.11), and (1.12).

The gate capacitance changes depending on whether the transistor is biased in the (a) accumulation region, (b) off region, (c) linear region, or (d) saturation region, as shown in Fig. 1.12. Within each region, the gate-source capacitance C_{GS}, the gate-drain capacitance C_{GD}, and the gate-body capacitance C_{GB} all behave differently.

In the accumulation region (a), a negative potential is applied to the NMOS gate. Thus, positive charge accumulates where a channel of electrons would form in the regular ON state, and the gate-body capacitance is formed.

In the off region (b), the gate-source and gate-drain capacitances C_{GS0} and C_{GD0} can be attributed to the gate and source/drain overlap capacitance C_1, the gate side edge and source/drain fringe capacitance C_2, and gate and source/drain side edge

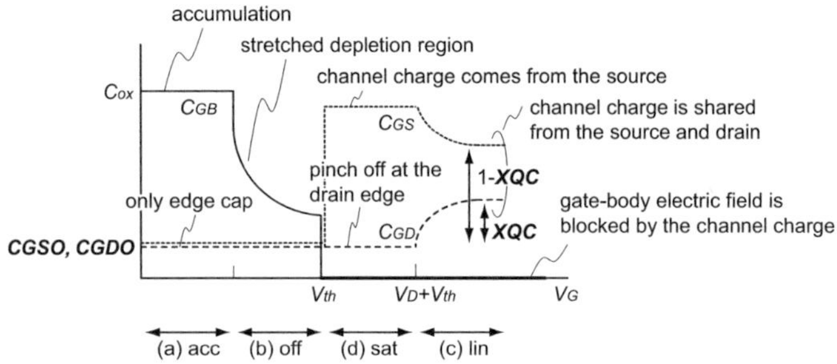

Fig. 1.13 Gate capacitance as a function of gate voltage

fringe capacitance C_3. These are collected and treated as single parameters **CGSO** and **CGDO**. Also, the gate-body capacitance varies as the width of the depletion layer changes, and as the gate potential is increased, the depletion region width increases and therefore the capacitance decreases.

When the channel is formed in the transistor as in the linear region (c), capacitance with the channel C_4 is created, shielding the fringe capacitance with the source side edge C_3. Here, the channel charge is injected from either the source or the drain, and the gate-body capacitance becomes zero, while the gate-source and gate-drain capacitances are formed. The ratio of charge injected by the source:drain to form the channel is expressed as source:drain = $1 - \mathbf{XQC} : \mathbf{XQC}$, and empirically, it is said that $\mathbf{XQC} = 0.4$.

In the saturation region (d), the channel is pinched off at the drain side, decreasing the gate-drain capacitance.

The gate-body/source/drain capacitance is plotted in Fig. 1.13. The horizontal axis is the gate voltage, and as the gate voltage is increased from some negative value, the regions change from (a) accumulation region → (b) off region → (d) saturation region → (c) linear region. If instead the horizontal axis was the drain voltage, then the regions would change from (c) linear region → (d) saturation region.

We have discussed the gate capacitance in great detail, but it is important not to forget the voltage-dependent PN junction capacitance of the source/drain-body modeled by Eqs. (1.10), (1.11), and (1.12), and shown in Fig. 1.7.

These changes in the transistor capacitances are built into SPICE model equations, and the SPICE simulator calculates them even without the circuit designer realizing this, but it is important to understand the physical phenomena such as these capacitances occurring within the transistor.

For example, in the inverters shown in Fig. 1.14a, the second inverter as the load of the first inverter is often represented as a single capacitance C_L terminated to G_{ND} as shown in Fig. 1.14b. However, reality looks closer to Fig. 1.14c, where not only are there capacitances terminated to G_{ND}, but also there are capacitances terminated to the supply as well as those connected serially. Furthermore, these capacitances change in time with the applied voltage. Figure 1.14b is an immense

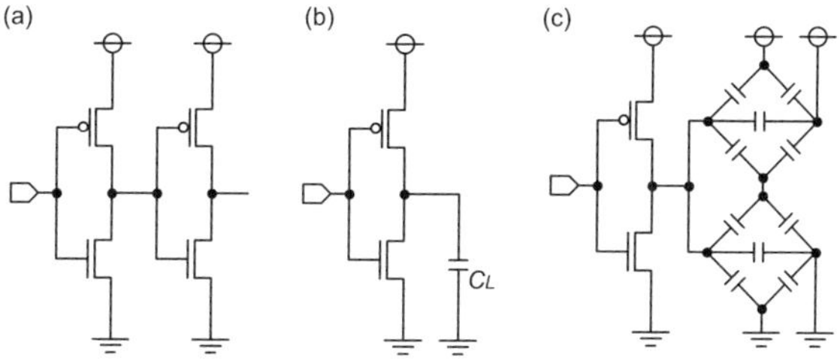

Fig. 1.14 Load capacitance

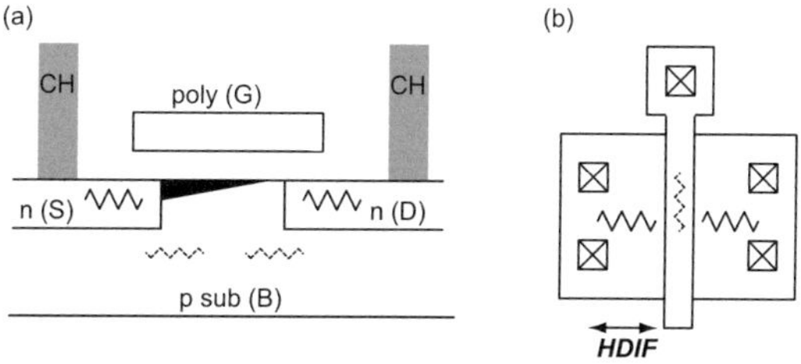

Fig. 1.15 Source and drain resistances

simplification of this, and the model of Fig. 1.14c is required when considering the detailed behavior of charge and current.

1.2.2.4 Parasitic Resistances of the Transistor

The source/drain is connected to metal wires by contact holes (CH), and there is a parasitic resistance from the contact to the gate edge, which must be modeled (Fig. 1.15). Old models (LEVEL1~3) define parameters **RD** and **RS** that are independent of W and **HDIF**, while models like BSIM1 and BSIM2 include these resistances in the drain current equations. In BSIM3, parameters for the resistance per unit W **RDSW**, gate-body bias dependence parameters **PRWG** and **PRWB**, and a temperature dependence parameter **PRT** are defined, and the parasitic resistance is calculated as

$$R_{ds} = \frac{\left[\mathbf{RDSW} + \mathbf{PRT} \cdot \left(\frac{T}{T_{\mathrm{nom}}} - 1\right)\right] \cdot \left\{1 + \mathbf{PRWG} \cdot V_{gsx} + \mathbf{PRWB} \cdot \left[(\phi_s - V_{bsx})^{\frac{1}{2}} - \phi_s^{\frac{1}{2}}\right]\right\}}{W^{\mathbf{WR}}}. \tag{1.18}$$

In addition, because BSIM3 does not include the gate resistance and the body resistance, these must be added if needed as separate resistances in the schematic diagram for simulation.

1.2.3 SPICE Parameters

1.2.3.1 SPICE Parameter File

In circuit design, it is possible to specify which models and parameters to use for each transistor. However, in general, a process engineer who designs the physical transistor structure provides the model and parameters as a "SPICE parameter file," which the circuit designer in turn applies to all transistors in the design.

The contents of a SPICE parameter file, for example, for HSPICE, look as follows:

```
.MODEL NLP NMOS
+ LEVEL = 53
+ VERSION = 3.2
+ TOX = 10e-9
+ U0 = 300
+ CJ = 0.1
+ CJSW = 0.001
+ MJ = 0.667
+ ....

.MODEL NHP NMOS
+ LEVEL = 53
+ VERSION = 3.2
+ TOX = 9e-9
+ U0 = 330
+ CJ = 0.1
+ CJSW = 0.001
+ MJ = 0.667
+ ....
```

`.MODEL NLP NMOS` and `.MODEL NHP NMOS` define two types of models and sets of parameters for the NMOS, `NLP` and `NHP`. Both are `LEVEL=53 VERSION=3.2`, which specify the BSIM3v3.2 "transistor model" that defines the relationships between the voltages, currents, capacitances, and resistances. The rest specify parameter values used in the transistor model, such as the gate oxide thickness, the mobility, and the drain-body junction capacitance per unit area. It is important to note that inside this `.MODEL`, both the transistor "model," which specifies the equations describing the voltages and currents, and the "parameter values," which are used within the model, are specified simultaneously. These are often combined and simply called the "model" or the "parameters," which is a source of confusion. The original meaning of the terms "model" and "parameter" must be understood to recognize what is being referred to. Within this text, we refer to the name specified by `.MODEL` (e.g., `NLP, NHP`) as the "model name" or the "model," each variable (e.g., `TOX, U0`) as "model parameters," models and parameters together as "SPICE parameters," and equations describing the current and voltage relationships as "transistor models." Thus, we would say, for example, "The model name specified by this SPICE parameter is `NLP`, and the value of one of the model parameters `TOX` is 10 nm. From the model parameter value of

`LEVEL=53, VERSION=3.2`, we know that the transistor model being used is BSIM3v3.2."

During circuit design, the designer specifies the model name to use, such as `NLP` and `NHP`, for each transistor. That is, in Sect. 1.1.2, the model is specified by the model name. For example, `NLP` would be used for a circuit requiring low power consumption, `NHP` would be used for a circuit requiring high operating speed, and so on.

Before starting the circuit design, take a look at the SPICE parameter file, find which model names (such as `NLP`) are available, understand each of their features (such as high performance or low power), and check which transistor model (such as BSIM3 or BSIM4) is being used.

1.2.3.2 Size-Dependent Applicable Range of Model

Depending on the transistor size, physical phenomena may or may not be significant enough to be observed. Therefore, model parameter values may be changed based on the transistor size. An ideal transistor model and ideal model parameter values would not have this issue, but practical transistor models are approximations of reality as are the model parameter values, leading to this inconvenience.

For example, with the SPICE parameters as follows:

```
.MODEL NLP.1 NMOS
+ LMIN = 0.18e-6
+ LMAX = 0.30e-6
+ WMIN = 0.50e-6
+ WMAX = 5.00e-6
+ LEVEL = 53
+ VERSION = 3.2
+ TOX = 10e-9
+ U0 = 300
+ CJ = 0.1
+ CJSW = 0.001
+ MJ = 0.667
+ ....

.MODEL NLP.2 NMOS
+ LMIN = 0.30e-6
+ LMAX = 0.60e-6
+ WMIN = 0.50e-6
+ WMAX = 5.00e-6
+ LEVEL = 53
+ VERSION = 3.2
+ TOX = 10e-9
+ U0 = 300
+ CJ = 0.1
+ CJSW = 0.001
+ MJ = 0.6
+ ....
```

transistors in the range $0.18\,\mu\text{m} < L < 0.30$ and $0.5\,\mu\text{m} < W < 5.0\,\mu\text{m}$ would use `NLP.1`, and transistors in the range $0.30\,\mu\text{m} < L < 0.60\,\mu\text{m}$, $0.5\,\mu\text{m} < W < 5.0\,\mu\text{m}$ would use `NLP.2`. In this way, multiple models can account for a wide range of L and W. It is not necessary to specify the model name of each transistor on the schematic as `.1`, for example. Entering `NLP` suffices and the simulator will read in the appropriate model based on the L and W, and use that model to run simulations.

1.2.3.3 Process Corner Libraries

Due to manufacturing variations, transistor characteristics may deviate from the target. For example, the gate oxide thickness may vary during fabrication. For the SPICE parameters listed below, while the typical (`T`) thickness is 10 nm, this value can potentially vary within the range from 9 to 11 nm. When the gate oxide thickness is 11 nm, the transistor characteristics are degraded and the circuit operating speed is slow (`S`), whereas at 9 nm the circuit becomes fast (`F`), and these process "corners" are indicated by the subscripts `T`, `S`, and `F`.

However, specifying which process corner to use as the simulation condition is entered elsewhere and not done on the circuit schematic, where all that is necessary is to specify the model name, for example, as `NLP`.

```
.MODEL NLP NMOS
+ LEVEL = 53
+ VERSION = 3.2
+ TOX = tox
+ U0 = 300
+ CJ = 0.1
+ ....

.LIB NT
.PARAM tox = 10e-9
.ENDL

.LIB NS
.PARAM tox = 11e-9
.ENDL

.LIB NF
.PARAM tox =  9e-9
.ENDL
```

1.2.4 Understanding the Model

In the process of designing a transistor circuit, specifying the model, and simulating the circuit behavior, you need to be constantly aware of whether some physical phenomenon is reflected in the model or not and if not how to model this behavior in circuit simulations. This may not be necessary when designing relatively

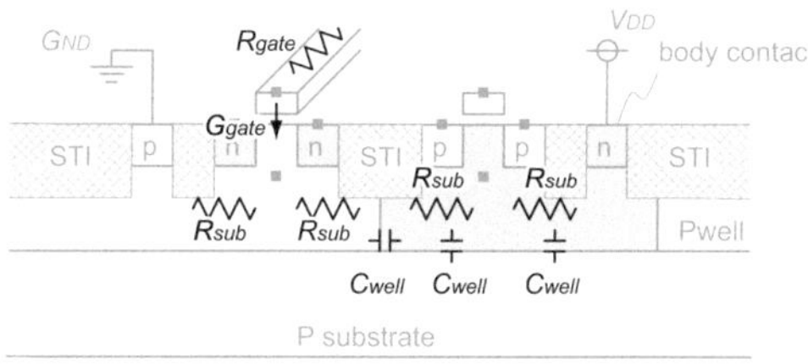

Fig. 1.16 Subtle physical phenomena, which may or may not be modeled

straightforward circuits, but in unusual situations such as changing the body bias or simulating the details of leakage currents, it is imperative that you know whether the internal physical phenomena of the circuit are modeled or not. For example, the subthreshold current of a LEVEL1 model transistor is zero, which means this model cannot be used to design circuits in the subthreshold regime. Some points to keep in mind are listed (refer to Fig. 1.16):

- The capacitance between the substrate and the well C_{well} is not accounted for. Thus, obviously, the deep N-well structure is also not included. Be careful when dynamically controlling the body bias.
- The resistance between the body and well contact is not accounted for and is thus zero. Be careful if you are considering supply noise and/or substrate noise.
- The gate poly resistance R_{gate} is zero in BSIM3, but is taken into account in BSIM4. Take care especially with high-frequency circuitry.
- Gate leakage conductance G_{gate} is assumed to be zero in BSIM3, but is taken into account in BSIM4. Be careful when designing low-power circuits or simulating leakage currents.
- BSIM3 will automatically calculate AD/AS/PD/PS based on the parameter **HDIF**. On the other hand, **HDIF** is invalid in BSIM4, and unless AD/AS/PD/PS are specified, they will be evaluated as zero. Be wary of whether the schematic editor supports AD/AS/PD/PS generation.

1.3 Techniques for Circuit Design

1.3.1 Hierarchical Design

It is important to always keep hierarchical design in mind when going about circuit design. For example, when designing an inverter chain, do not draw the circuit as in Fig. 1.17a. Rather, create a single inverter cell and then arrange those cells as in Fig. 1.17b. When the inverter sizes need to be changed, the design in Fig. 1.17a would require all transistor sizes to be changed one by one, whereas the design in Fig. 1.17b would require only a change in the inverter cell, which would get reflected in all instantiated inverters. Also, this method makes the layout and its verification easier later on.

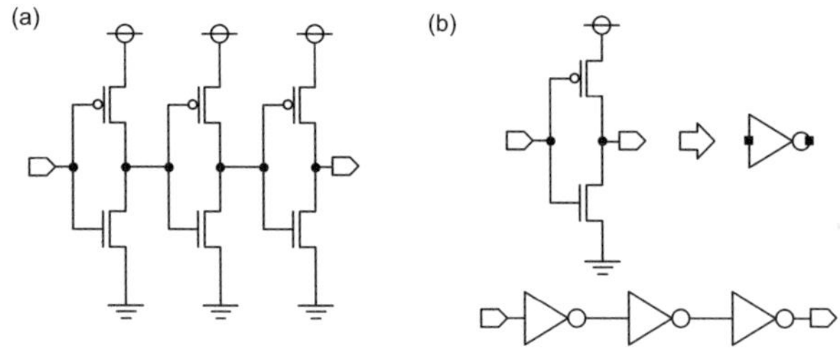

Fig. 1.17 Hierarchical design: (**a**) flat, (**b**) hierarchical [recommended]

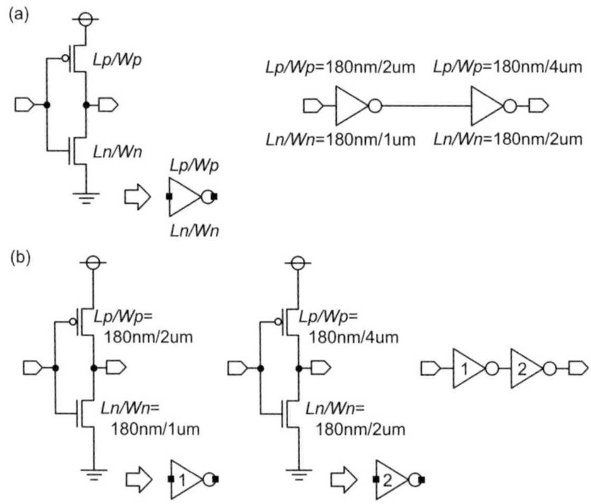

Fig. 1.18 Cell parameterization. (**a**) Parameterized, (**b**) different cells [recommended]

It is also possible to give parameters to the cells, but this is not recommended due to the confusion this may cause later. For example, the inverter's $L_p/L_n/W_p/W_n$ can be parameterized as in Fig. 1.18a and these parameters can be set for each instantiated symbol. However, this will become a pain in the layout stage because the schematic and layout will not correspond. Instead, creating a different cell for each type of inverter as in Fig. 1.18b will allow for a smooth design flow.

1.3.2 Dealing with Supply and Ground

Care must be taken during circuit design when handling the power supply and ground terminals. One method is to define the supply and ground as global variables, visible from anywhere in the circuit (global variables in Cadence Virtuoso are defined as `Vdd!, Gnd!`, with exclamation marks). In Fig. 1.17b, supply and

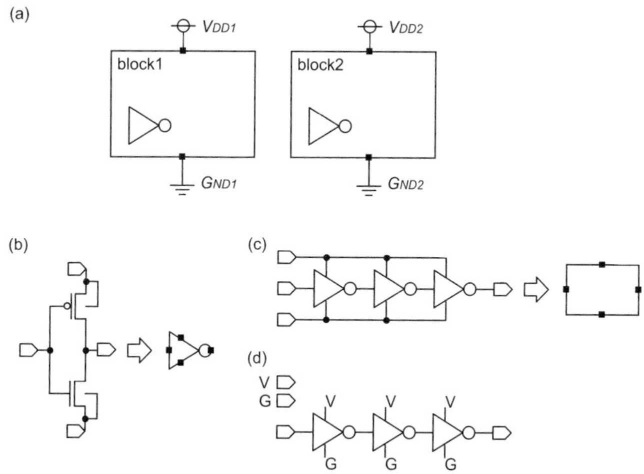

Fig. 1.19 Power and ground terminals

ground are assumed to be global and visible from anywhere and therefore are not defined as terminals when creating the symbol.

However, as the design progresses, the need to separate voltage domains will come up, as in Fig. 1.19. Separating the supplies of the internal circuits and the input-output buffers and separating the digital and analog supplies are typical examples. These scenarios cannot be handled if the supply and ground within the inverter cell are set to be global by using `Vdd!`, `Gnd!`. Therefore, it is strongly recommended that the supply and ground are treated as terminals as in Fig. 1.19b, rather than using global variables. However, this method requires that the supply and ground terminals of all symbols be connected as in Fig. 1.19c. To avoid a messy circuit diagram, it is possible to assign supply and ground labels to nodes, which will connect all nodes with the same label as shown in Fig. 1.19d (most schematic editors have this feature).

1.3.3 Connections by Labeling

It is suggested that this connection by labeling is not used anywhere except for supply and ground connections. For example, by using labels, the circuit in Fig. 1.20a becomes that shown in Fig. 1.20b. The latter circuit appears cleaner at first sight, but it will become apparent that this circuit is more difficult to follow when the circuit is being analyzed by observing waveforms or the design is at the layout stage. Although there are a few scenarios where the label connection strategy may seem useful, such as the CLK distribution to DFFs, in general it is recommended that the label connection method be applied only to supply and ground.

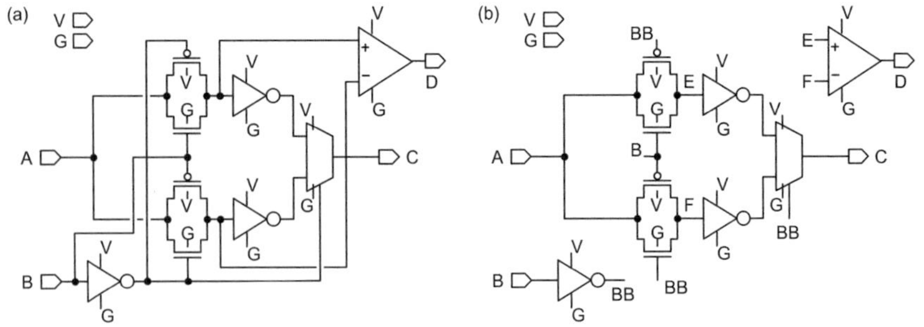

Fig. 1.20 Label connection. (a) Power and ground only [recommended], (b) label connection for signal wires

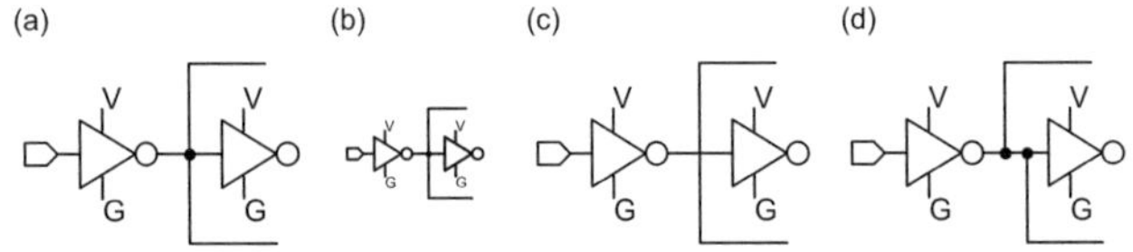

Fig. 1.21 A connection point

1.3.4 Connection Points

Cross connections on circuit diagrams, such as the one shown in Fig. 1.21a, should be avoided. Especially when the schematic is zoomed out as in Fig. 1.21b, it becomes difficult to distinguish between this and a set of unconnected wires (Fig. 1.21c). Make sure to shift the connections slightly and make the connections at different points, as shown in Fig. 1.21d.

Chapter 2
SPICE Simulation

After the circuit schematic is drawn, the next step is to simulate it in SPICE. "So, what exactly *is* a SPICE simulation?"

2.1 Principles of Simulation

SPICE (Simulation Program with Integrated Circuit Emphasis) simulations are the most basic simulations in LSI design. SPICE simulations can be categorized into DC analysis, AC analysis, transient (TRAN) analysis, and harmonic balance (HB) analysis. The design is conducted by using suitable simulations, based on the understanding of their characteristics and principles.

2.1.1 *DC Analysis*

DC analysis determines the final voltages of the circuit, while the input voltages are held constant. Because the changes in voltages with time are ignored, capacitances can be thought of as open circuits (0[F]) and inductances as shorts (0[H]), as shown in Fig. 2.1a, b. Therefore, the circuit shown in Fig. 2.1c can be treated as the circuit in Fig. 2.1d.

2.1.2 *Linear Circuit Elements*

What will the node voltages and branch currents of the circuit in Fig. 2.1d be like? SPICE internally uses equations of the form $I = GV$ rather than $V = RI$ for

T. Nakura, *Essential Knowledge for Transistor-Level LSI Circuit Design*,
DOI 10.1007/978-981-10-0424-7_2

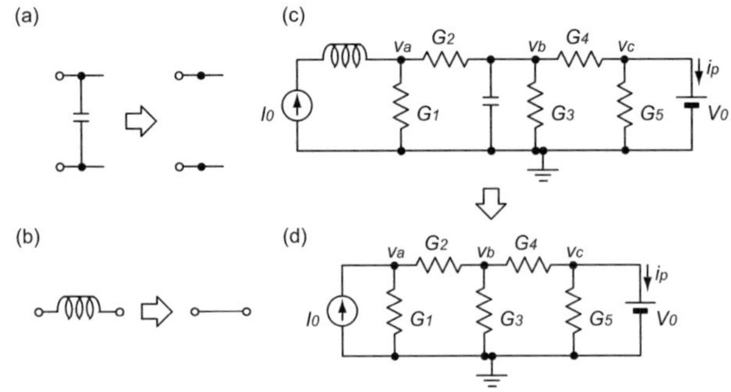

Fig. 2.1 DC analysis

calculations. According to current continuity equations (Kirchhoff's Current Law (KCL)) applied to the currents at node a:

$$-I_0 + G_1 v_a + G_2(v_a - v_b) = 0, \tag{2.1}$$

and similarly, at node b and node c:

$$G_2(v_b - v_a) + G_3 v_b + G_4(v_b - v_c) = 0, \tag{2.2}$$

$$G_4(v_c - v_b) + G_5 v_c + i_p = 0. \tag{2.3}$$

Additionally,

$$v_c = V_0. \tag{2.4}$$

These equations can be expressed in matrix form:

$$\left(\begin{array}{ccc|c} G_1+G_2 & -G_2 & 0 & 0 \\ -G_2 & G_2+G_3+G_4 & -G_4 & 0 \\ 0 & -G_4 & G_4+G_5 & 1 \\ \hline 0 & 0 & 1 & 0 \end{array}\right) \begin{pmatrix} v_a \\ v_b \\ v_c \\ \hline i_p \end{pmatrix} = \begin{pmatrix} I_0 \\ 0 \\ 0 \\ \hline V_0 \end{pmatrix} \tag{2.5}$$

More generally,

$$\begin{pmatrix} \boldsymbol{G} & \boldsymbol{F} \\ \boldsymbol{B} & \boldsymbol{R} \end{pmatrix} \begin{pmatrix} \boldsymbol{v} \\ \boldsymbol{i} \end{pmatrix} = \begin{pmatrix} \boldsymbol{C} \\ \boldsymbol{E} \end{pmatrix} \tag{2.6}$$

where $\boldsymbol{G}$ is the conductance matrix, $\boldsymbol{C}$ and $\boldsymbol{E}$ are the current and voltage sources, $\boldsymbol{v}$ represents the node voltages, and $\boldsymbol{i}$ represents the current flowing into the voltage source. By solving these equations, each node voltage v_x and the current flowing into the voltage source i_p can be found (in the world of simulators, it is customary to write the voltage/current sources, which become the boundary conditions, as uppercase letters, and the node voltages/currents, which are the variables, as lowercase letters).

The diagonal elements g_{ii} of the conductance matrix $\boldsymbol{G}$ are the sum of all branch conductances connected to the ith node, and the elements g_{ij} outside of the diagonal are the sum of all branch conductances that connect nodes i and j, with a minus sign. Current sources come to the right-hand side, and the values are positive if the current flows into the node and negative if it flows out (in this example, the node from which the current flows out is the reference node and, thus, does not appear in the equations). Furthermore, in the presence of voltage sources, currents flowing into the source are defined as the positive direction, and extra rows and columns are added.

The left-hand side of the equation would have been a 3×3 admittance matrix $\times$ voltage variables and the right-hand side the current sources only, but because of the existence of the voltage source, the matrix is revised to become a 4×4 matrix. This method of setting up the circuit equations is called modified nodal analysis.

The point here is that as long as the circuit impedances and the connection information are known, this matrix can be constructed automatically. Also, the voltages and currents at each node can be obtained by solving this equation (inverting the matrix). Because the waveforms can be automatically calculated without having to analyze the circuit, it can be said that this is a general simulation method.

2.1.3 Nonlinear Circuit Elements

When the circuit elements are linear, the relationship $I = GV$ always stands, meaning Eq. (2.5) also holds and can be solved to determine the currents and voltages. What should we do in the nonlinear situation? For example, in Fig. 2.2a, what is the output voltage v_o when $V_{DD} = 1$ V and $V_{IN} = 0.3$ V?

Because the transistor model is specified, the drain current dependencies on the terminal voltages are known. That is, the function $I_D = f(V_G, V_D, V_S, V_B)$ is known, and the partial derivatives $\frac{\partial I_D}{\partial V_G}$, $\frac{\partial I_D}{\partial V_D}$, $\frac{\partial I_D}{\partial V_S}$, and $\frac{\partial I_D}{\partial V_B}$ are also known (if the analytical formulae are given, it is possible to analytically take their partial derivatives. Usually in SPICE, the partial derivative equations are given as a part of the model

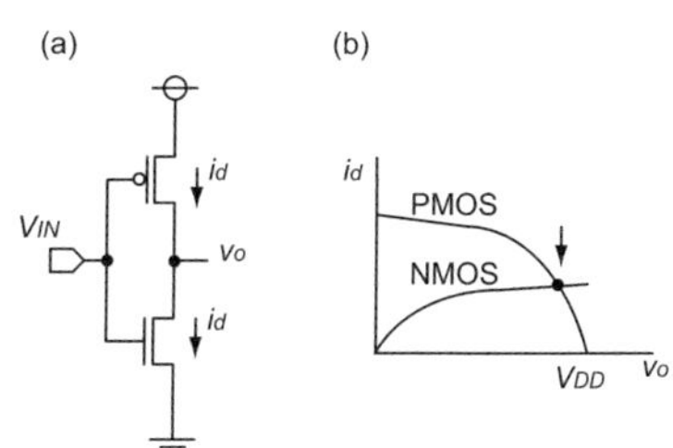

Fig. 2.2 Operating point for nonlinear elements

equations. However, even if the analytical expressions for the partial derivative are not available, numerical values can be calculated by evaluating, for example:

$$\left.\frac{\partial I_D}{\partial V_G}\right|_{V_{G0},V_{D0},V_{S0},V_{B0}} = \frac{I_D(V_{G0}+\Delta V_G, V_{D0}, V_{S0}, V_{D0}) - I_D(V_{G0}, V_{D0}, V_{S0}, V_{D0})}{\Delta V_G} \tag{2.7}$$

using the equation for I_D).

In Fig. 2.2a, supposing the PMOS and NMOS currents as a function of v_o are as plotted in Fig. 2.2b because the V_G, V_S, and V_B of the PMOS and NMOS are fixed, then v_o is the cross point of the two curves. This is evident from the current continuity equation at the output which states that the currents flowing in the PMOS and NMOS are equal. This intersection point can be determined by solving the equations given by the analytical expressions of the model. For example, if the drain current equations are given as Eqs. (1.4) and (1.5), the second-order equations can probably be solved analytically. However it is much too complicated to analytically solve the equations given by BSIM. SPICE utilizes the full powers of its calculation abilities to arrive at numerical results.

If the model equations are given, then based on some assumed voltages for each node (D, G, S, B), the current I_D and its partial derivative $\frac{\partial I_D}{\partial V_x}$ can be calculated. In this example, the boundary conditions are given as $V_{IN} = 0.3\,\text{V}$ and $V_{DD} = 1\,\text{V}$. First, the output voltage v_o is assumed to be some value v_k as an initial guess (in SPICE this value is often zero). The values of the currents i_{pk} and i_{nk} that flow through the PMOS and NMOS when the output voltage is equal to v_k can be calculated numerically, and the value of the derivatives at that point $g_{pk} \equiv \left.\frac{\partial i_{dp}}{\partial v_o}\right|_{v_o=v_k}$ and $g_{nk} \equiv \left.\frac{\partial i_{dn}}{\partial v_o}\right|_{v_o=v_k}$ can also be calculated. As shown in Fig. 2.3a, by using the calculated PMOS and NMOS current values, the equations $i_p = g_{pk}(v - v_k) + i_{pk}$ and $i_n = g_{nk}(v - v_k) + i_{nk}$ can be written, from linear approximations around the voltage v_k. This is in fact the same as substituting Fig. 2.3b for Fig. 2.2a. The new circuit consists only of voltage sources, current sources, and resistors, which means that the current and voltage values, i.e., the value at point P_{k+1} in Fig. 2.3a,

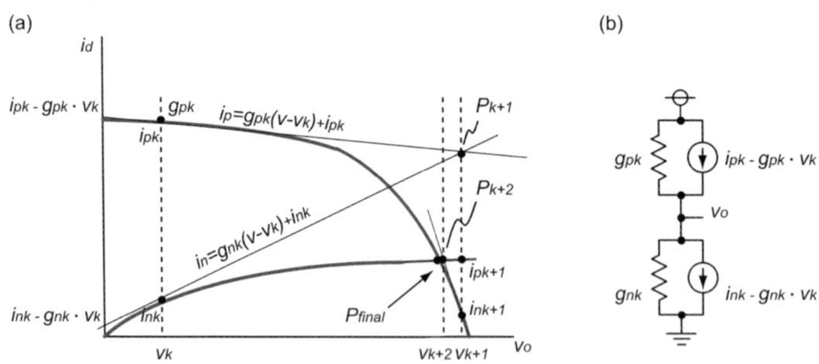

Fig. 2.3 Solving for the operating point of nonlinear elements

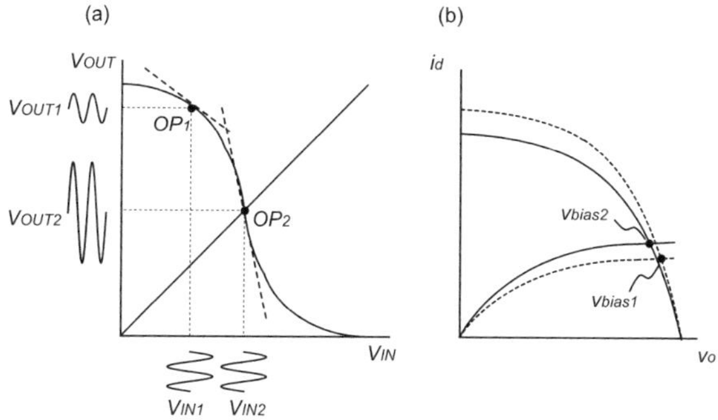

Fig. 2.4 The input-output characteristics of an inverter

can be easily determined through modified nodal analysis. The current equations for the PMOS and NMOS are again linearized around v_{k+1}, and a new point P_{k+2} is determined through modified nodal analysis. After repeating this process, when P_{n+1} has barely moved with respect to P_n, that is the final value of convergence v_{final}, which is our calculated value for the output voltage of the circuit. This technique is called Newton's method or the Newton-Raphson (NR) method and is a common method for solving nonlinear equations.

The points here are that the value for a current for a particular voltage, as well as the partial derivative value for the current, can be calculated given the model equations and that linearization around guess points is repeated until the value converges to our destination. Therefore, while the matrix equation needed to be solved only once for a circuit of linear elements, iterative calculations are required for nonlinear circuits.

In addition, when determining the input-output characteristics of an inverter, like the solid line in Fig. 2.4a, we search for the output voltage by varying the input voltage V_{IN} little by little, as in Fig. 2.4b. For example, after the output voltage at $V_{IN} = 0.3$ V has been determined, to evaluate the output voltage at $V_{IN} = 0.31$ V, the initial guess can be set to the solution of $V_{IN} = 0.3$ V to arrive at the final point in fewer iterations.

2.1.4 AC Analysis

If the inverter input voltage was $V_{IN} = 0.3 + 0.1 \sin(2\pi ft)$, what would the output voltage be? In AC analysis, DC analysis (also called operating point (OP) analysis) is first conducted to determine the voltage of each node at $V_{IN} = 0.3$ V. Then, the differential impedance of each element at and around this bias point is calculated to determine the circuit characteristics. This is in fact a linear approximation of the circuit characteristics around the bias point, as shown in dotted lines in Fig. 2.4a. Therefore, the gain from the input to the output is small around OP_1 and large around OP_2. An important point is that AC analysis is a linear approximation

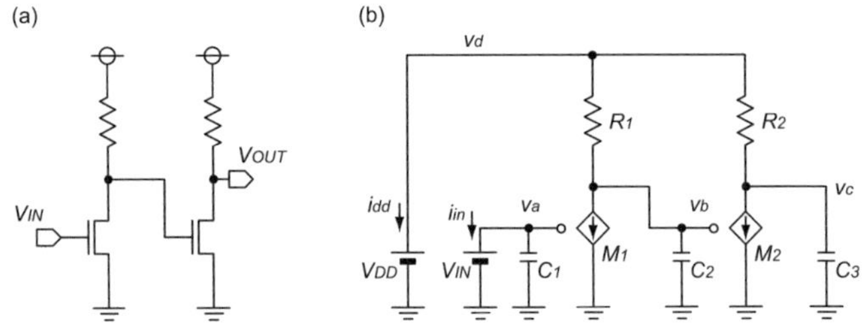

Fig. 2.5 The equivalent circuit of an inverter chain

and does not follow the input-output characteristic curve. For example, if V_{IN} of amplitude $V_{DD}/2$ is applied at OP_2, the output voltage given by AC analysis would be a value larger than V_{DD}. Also, the gain $v_{\mathrm{out}}/v_{\mathrm{in}}$ evaluated by AC analysis depends on the bias point but not the input amplitude (the small signal gain of AC analysis is represented with lowercase letters. Do not confuse this with the lowercase letters representing node voltages).

How are frequency dependencies taken into account? As we saw in Sect. 1.2.2, transistor capacitances depend on the bias voltages. That is, the capacitance value at the bias point is determined and the impedance is calculated as $1/j\omega C$.

The circuit in Fig. 2.5a can be represented as the equivalent circuit in Fig. 2.5b. Transistor capacitances are lumped into representative capacitances $C_1 \sim C_3$ in the figure, for simplicity (as we learned in Sect. 1.2.2, an actual transistor contains five capacitances). The circuit can be expressed as a matrix equation using modified nodal analysis:

$$\left(\begin{array}{cccc|cc} j\omega C_1 & 0 & 0 & 0 & 1 & 0 \\ g_{m1} & G_1 + j\omega C_2 + g_{d1} & 0 & -G_1 & 0 & 0 \\ 0 & g_{m2} & G_2 + j\omega C_3 + g_{d2} & -G_2 & 0 & 0 \\ 0 & -G_1 & -G_2 & G_1 + G_2 & 0 & 1 \\ \hline 1 & 0 & 0 & 0 & 0 & 0 \\ 0 & 0 & 0 & 1 & 0 & 0 \end{array}\right) \left(\begin{array}{c} v_a \\ v_b \\ v_c \\ v_d \\ \hline i_{\mathrm{in}} \\ i_{dd} \end{array}\right) = \left(\begin{array}{c} 0 \\ 0 \\ 0 \\ 0 \\ \hline V_{IN} \\ V_{DD}. \end{array}\right) \tag{2.8}$$

Here, we have defined the following:

$$g_m = \frac{\partial I_D}{\partial V_G}, \qquad g_d = \frac{\partial I_D}{\partial V_D}. \tag{2.9}$$

(Transistors are thought of as voltage-controlled current sources here. For example, the current $g_{m1}v_a$ is added as a current flowing out of node b, which makes the matrix element at index (1,2) equal to g_{m1}. This is not particular to AC analysis but also applies to DC and transient analyses as well). AC analysis linearizes the circuit under simulation around the bias point, so the solution to this matrix equation is the result of the AC simulation. Iterative calculations for inclusion of nonlinearity

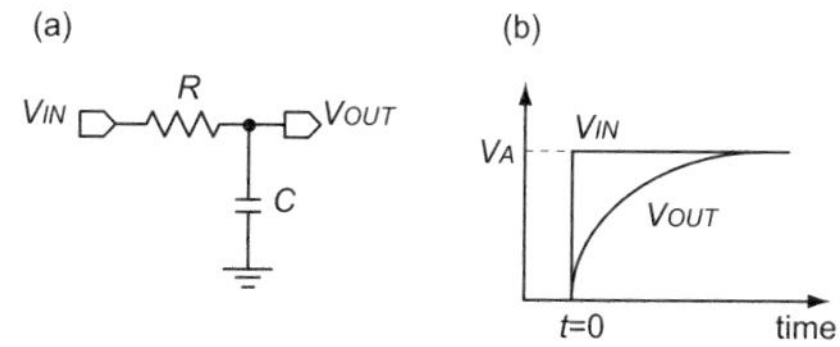

Fig. 2.6 An RC circuit and its transient response

is not conducted. As for the voltage source (V_{DD}, V_{IN}), AC voltages are used. A zero AC voltage is used on a constant voltage source such as the power supply. The constant voltage affects only for the calculation of the bias point where the circuit is linealized. In addition, frequency dependencies can be simply calculated by changing the value of ω.

2.1.5 Transient Analysis

The transient (TRAN) analysis solves for the time response waveform. As a simple example, let us solve for the transient response of the RC circuit shown in Fig. 2.6a.

2.1.5.1 Analytical Solution

If we let the current flowing to be I:

$$V_R + V_C = RI + \frac{1}{C}\int Idt = V_A \tag{2.10}$$

which we can differentiate to get

$$R\frac{dI}{dt} + \frac{1}{C}I = 0. \tag{2.11}$$

After separating the variables, we get

$$\frac{dI}{I} = -\frac{1}{RC}dt. \tag{2.12}$$

Integrate both sides to get

$$\log_e I = -\frac{1}{RC}t + K \quad , \tag{2.13}$$

or

$$I(t) = ke^{-\frac{1}{RC}t} \quad (k = e^K) \tag{2.14}$$

where K is the constant of integration. If the charge on the capacitance is zero at time $t = 0$,

$$I(t = 0) = \frac{V_A}{R} \tag{2.15}$$

from which we can say that the current is

$$I(t) = \frac{V_A}{R} e^{-\frac{1}{RC}t} \tag{2.16}$$

and the output voltage is

$$V(t) = V_A - RI(t) \tag{2.17}$$

$$= V_A \left\{ 1 - e^{-\frac{1}{RC}t} \right\}. \tag{2.18}$$

Here we have solved differential equations, and we would arrive at the same answer if we used Laplace transforms.

As you can imagine, this method of analytically determining the waveform by solving differential equations cannot be used to solve circuit equations which contain transistors that have more complex analytical equations. Are there other methods?

2.1.5.2 Forward Euler Method

We had to solve differential equations for the transient analysis of the RC circuit because the value of the current $I(t)$ changed with time due to charge accumulating on the capacitor. Then, what happens if we divide time into small segments h and assume that

$$v_{n+1} - v_n = \frac{1}{C} i_n \times h, \tag{2.19}$$

that is, assume that within the small time window h, the current that flows is a constant. At time $t = 0$, the capacitance voltage is zero, which means

$$i_0 = \frac{V_A}{R}, \tag{2.20}$$

and at time $t = 1h$, the voltage on the capacitor v_1 is

$$v_1 = \frac{1}{C} i_0 \times h. \tag{2.21}$$

The voltage across the resistor is

$$V_A - v_1 \tag{2.22}$$

which means the current flowing is

$$i_1 = \frac{V_A - v_1}{R} \tag{2.23}$$

which are all the equations necessary to express the state at time $t = 1h$. To move on to the next time step, the capacitor voltage at $t = 2h$ is

$$v_2 = v_1 + \frac{1}{C} i_1 \times h \tag{2.24}$$

which means the state at $t = 2h$ can be calculated using the state at $t = 1h$.

By repeating this process, the state at any given time can be calculated. The point here is that the current does not change during the small time interval h, and therefore, the voltage across the resistor and the charge stored on the capacitor can be expressed as $R \times I$ and I/C, respectively. Thus, there is no need to use the nonlinear equations in the theoretical transient analysis expressions. From this, we can apply linear theory to make a simple and automatic numerical analysis program possible. This method is called the forward Euler method. Intuitively, this can be represented as in Fig. 2.7.

It is noted that

$$v_1 = \frac{1}{C} i_0 \times h \tag{2.25}$$

$$= V_A \frac{1}{RC} h \tag{2.26}$$

$$v_2 = v_1 + \frac{1}{C} i_1 \times h \tag{2.27}$$

$$= V_A \frac{1}{RC} h + V_A \frac{1 - \frac{1}{RC} h}{RC} h \tag{2.28}$$

$$= V_A \left\{ \frac{1}{RC} h + \left(\frac{1}{RC} h - (\frac{1}{RC} h)^2 \right) \right\}; \tag{2.29}$$

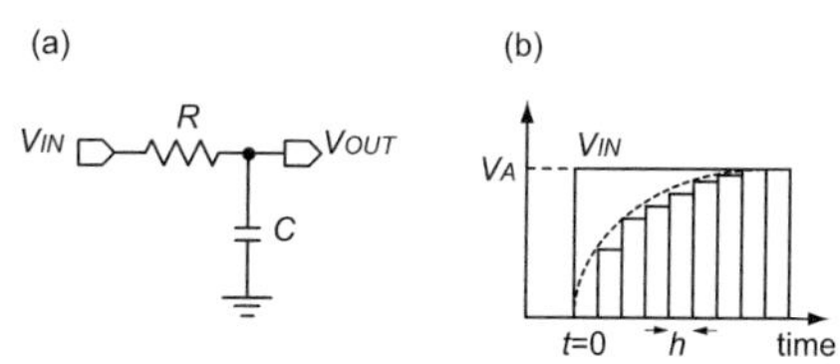

Fig. 2.7 Forward Euler method

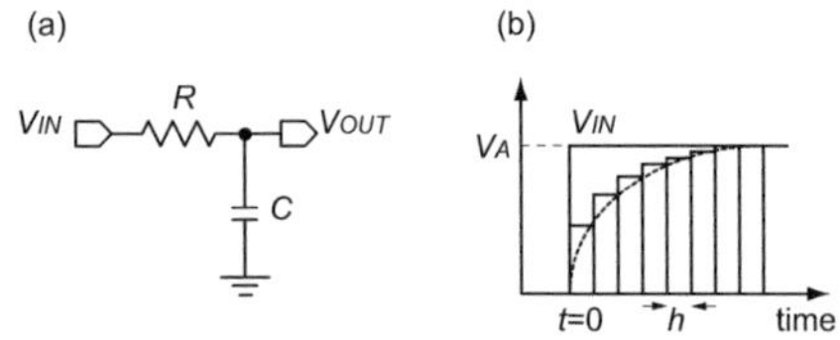

Fig. 2.8 Backward Euler method

thus, the rise in capacitance voltage slows down over time. When h is made infinitesimally small, the waveform matches that of Eq. (2.18).

2.1.5.3 Backward Euler Method

In the forward Euler method, we assumed that during time $nh \sim (n+1)h$, the current i_n that flows is determined by v_n. However, in the backward Euler method, we assume that the current i_{n+1} is determined by v_{n+1}. In other words,

$$v_{n+1} - v_n = \frac{1}{C} i_{n+1} \times h. \tag{2.30}$$

This seems slightly less intuitive, but it is known that this method yields a curve which is closer to the analytical solution than does the forward Euler method. The method can be thought of as depicted in Fig. 2.8.

This equation can be transformed as follows:

$$i_{n+1} = \frac{C}{h} v_{n+1} - \frac{C v_n}{h}. \tag{2.31}$$

When the state at t_n is known and we want to solve for the state at t_{n+1}, modified nodal analysis can be used by replacing the capacitance C with a parallel combination of a resistance $\frac{h}{C}$ $\left(\text{aconductance}\frac{C}{h}\right)$ and a current source $-\frac{Cv_n}{h}$.

2.1.5.4 Trapezoidal Method

The trapezoidal method has higher accuracy than the backward Euler method and is often utilized in SPICE. The trapezoidal method assumes that during the time interval $nh \sim (n+1)h$, the current that flows is the average of the current i_n, determined by v_n, and the current i_{n+1}, determined by v_{n+1}. In other words,

$$v_{n+1} - v_n = \frac{1}{C} \frac{i_{n+1} + i_n}{2} \times h. \tag{2.32}$$

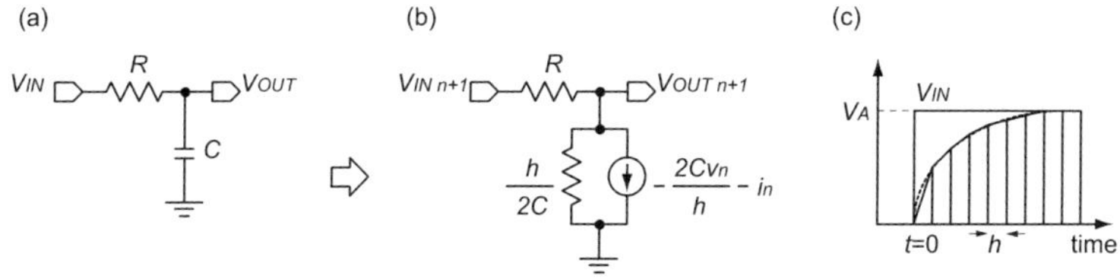

Fig. 2.9 Trapezoidal method equivalent circuit

This equation can be rewritten as

$$i_{n+1} = \frac{2C}{h}v_{n+1} - \frac{2Cv_n}{h} - i_n. \tag{2.33}$$

When the state at t_n is known and we want to solve for the state at t_{n+1}, modified nodal analysis can be used by replacing the capacitance C with a parallel combination of a resistance $\frac{h}{2C}$ (aconductance$\frac{2C}{h}$) and a current source $-\frac{2Cv_n}{h} - i_n$, as shown in Fig. 2.9a, b. This method can be visualized as depicted in Fig. 2.9c.

2.1.5.5 Transient Analysis for Nonlinear Circuits

What would happen if the capacitance values were not constant but rather dependent on the bias voltage? With the trapezoidal method, we can write

$$v_{n+1} - v_n = \frac{\dfrac{i_{n+1}}{C_{n+1}} + \dfrac{i_n}{C_n}}{2} \times h \tag{2.34}$$

which we can rewrite as

$$i_{n+1} = \frac{2C_{n+1}}{h}v_{n+1} - \frac{2C_{n+1}v_n}{h} - \frac{C_{n+1}}{C_n}i_n \tag{2.35}$$

(if we let $C_{n+1} = C_n = C$ in Eq. (2.35), this matches Eq. (2.33)). Thus, similar to what has been done in Fig. 2.9, the capacitor can be replaced by a parallel combination of an equivalent resistance $R_{eq} = \frac{h}{2C_{n+1}}$ and a current source $I_{eq} = -\frac{2C_{n+1}v_n}{h} - \frac{C_{n+1}}{C_n}i_n$. For example, if the capacitance is expressed as $C_{jo}\left(1 + \frac{v}{V_b}\right)^{-m}$, then we can write

$$G_{eq} = \frac{1}{R_{eq}} = \frac{2}{h}C_{jo}\left(1 + \frac{v_{n+1}}{V_b}\right)^{-m} \tag{2.36}$$

$$I_{eq} = -\left(\frac{2v_n}{h} + \frac{i_n}{C_n}\right)C_{jo}\left(1 + \frac{v_{n+1}}{V_b}\right)^{-m}. \tag{2.37}$$

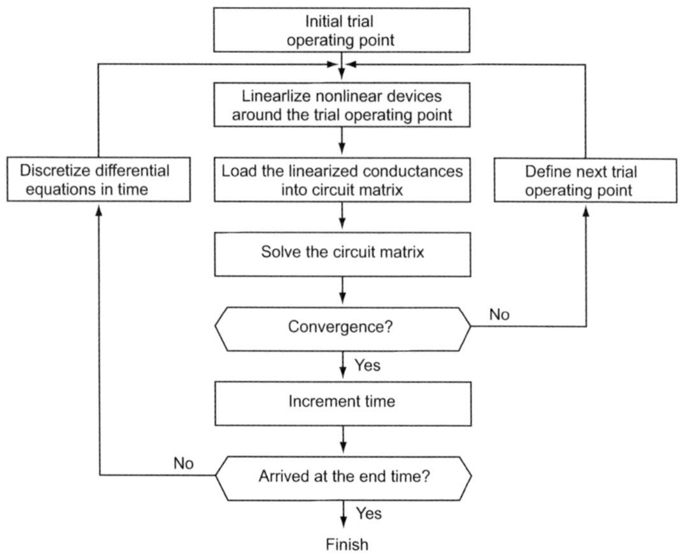

Fig. 2.10 Transient analysis simulation flow

Because v_n and i_n are known, the equivalent conductance G_{eq} as well as the equivalent current source I_{eq} can be thought of as nonlinear elements dependent on v_{n+1}. Thus, the voltages and currents at t_{n+1} can be iteratively calculated with modified nodal analysis which utilizes linearization by partial differentiation, which can be done with the Newton-Raphson method as was conducted in DC analysis.

2.1.5.6 Transient Analysis Flow

The above discussion was for the case of an R and a C, but even if the circuit contains nonlinear elements such as transistors, the result can be iteratively calculated by linearizing each time step. That is, voltage and current waveforms are calculated by following the flowchart in Fig. 2.10, which includes steps such as replacing the transistor with a voltage-controlled current source and the five equivalent capacitors, further replacing the voltage-controlled current source with a resistance (conductance) and a current source as in Fig. 2.3b, and replacing the capacitances with resistances (conductances) and current sources as in Fig. 2.9b.

2.1.6 Harmonic Balance Analysis

Harmonic balance (HB) analysis lies in between AC analysis and transient analysis. This is a simulation specialized for a representative frequency (f_0) and its higher-order frequencies (Nf_0). As with AC analysis, the waveforms of steady-state operation at that particular frequency are displayed, but unlike AC analysis, the nonlinearities of elements are also taken into account in the results.

This is often used in the simulation of RF circuits. For example, when signals of 1 GHz (f_1) and 999 MHz (f_2) are input to a mixer (multiplication) circuit, the difference of 1MHz appears as an output frequency component. To simulate this effect with transient analysis, time steps on the order of ps are required to observe the 1GHz waveform cleanly, but the simulation must be run for 1 μs to see one period of the output, which is time consuming. By using harmonic balance in situations like these, the magnitude and phase of each frequency component $Mf_1 + Nf_2$ can be calculated rapidly, including the circuit element nonlinearities.

2.1.7 Analysis Method Characteristics and Comparison

The output waveform of each analysis method, when a sinusoidal wave around the bias point (OP) is input to an inverter, is shown in Fig. 2.11. These differences must be well understood, and the optimal analysis method must be chosen during your circuit design.

2.1.7.1 DC Analysis

- Calculate the steady state point.
- Capacitors are open circuits, and inductors are shorts.
- Transistors follow the model equations accurately.
- Iterative calculations are necessary for simulation of nonlinear elements.

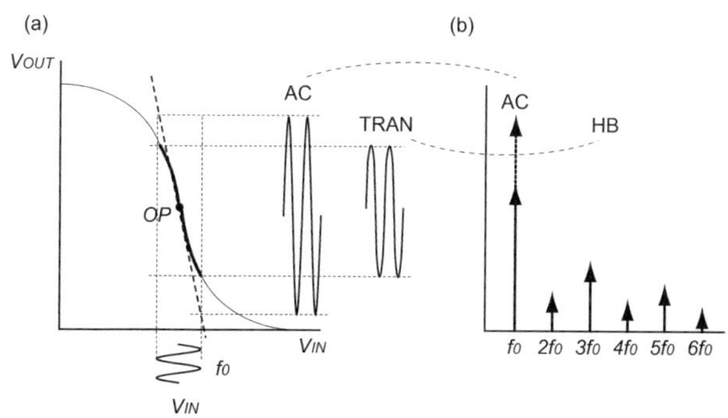

Fig. 2.11 Comparison of analysis methods

2.1.7.2 AC Analysis

- Calculate frequency dependencies.
- Find the bias point with DC analysis, and only consider the linearized characteristics around the bias point during AC analysis. Because it is assumed that there are no nonlinearities, the gain does not depend on the input voltage amplitude (but it does depend on the bias point).
- Capacitors are treated as impedances of value $1/j\omega C$.
- There are no nonlinearities because the circuit is linearized around the bias point. Therefore, there is no need for iterative calculations.
- The only frequency-dependent characteristics are that of capacitances (including the transistor PN junction capacitances) whose impedances are $1/j\omega C$.

2.1.7.3 TRAN Analysis

- This calculates the time response waveforms.
- The current is assumed to be constant (quasi-static) during each time step.
- To include nonlinear elements, iterative calculations at each time step are necessary.

2.1.7.4 HB Analysis

- The waveforms of the steady-state operation at a specified frequency are calculated by considering only the specified frequency and its integer multiple frequencies. The designer specifies the maximum order frequency to consider.

2.2 Fast SPICE

A simulator called Fast SPICE is used when we would like to conduct accelerated simulations, even at the cost of slightly degrading the accuracy. Synopsys NanoSim and Cadence Virtuoso UltraSim are typical examples. Some examples of accelerating methods are

- partitioning
- event-driven simulation
- time step control
- simplification of the model
- automatic decision and specification of simulation accuracy

and so on.

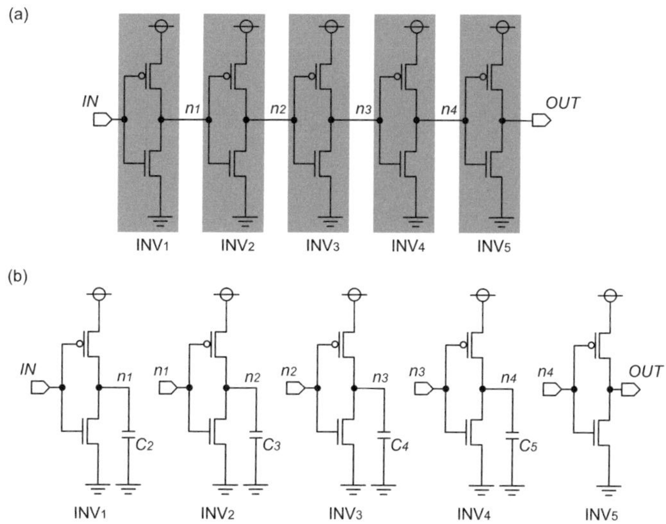

Fig. 2.12 Partitioning of a five-stage inverter chain circuit

2.2.1 *Partitioning and Event-Driven Simulation*

Let us consider the simulation of a five-stage inverter chain, as shown in Fig. 2.12a. When the voltage of *IN* changes, the voltage at n_1 will also change, but would the states of INV_3, INV_4, INV_5 affect the change of n_1? As we learned in the last chapter, the gate input capacitance of INV_2 depends on the voltage of n_2, because whether the transistors of INV_2 are in the linear region or the saturation region depends on the voltage of n_2. For the same reason, the change in voltage at n_2 depends on the voltage of n_3, which depends on the voltage at n_4, and so on. However, intuitively, we can see that the change in n_1 is probably not strongly affected by the change in n_3 and n_4. Most likely, a sufficient accuracy can be obtained if the gate input capacitance of INV_3 is set to the value C_3 at $V_{n_2} = V_{DD}/2$ and the simulation is run with only INV_1, INV_2, and a constant capacitance C_3. In some cases, simulating for the voltage change of n_1 with just INV_1 and a constant capacitance C_2 may give results with enough accuracy. Using this result for the voltage of n_1 as input, the voltage waveform of n_2 can be simulated with only INV_2 and C_3. The results of this can be used as the input waveform for the simulation with INV_3 and C_4, and so on. Thus, all of the node voltage waveforms from the input to the output of the five-stage inverter chain can be determined.

This division of the circuit is called partitioning. Also, in this example, the simulation of INV_2 is conducted only when n_1 changes, and the simulation of INV_3 is conducted after n_2 changes, and so on. This method of conducting simulations of partitioned blocks only when the input has changed is called event-driven simulation.

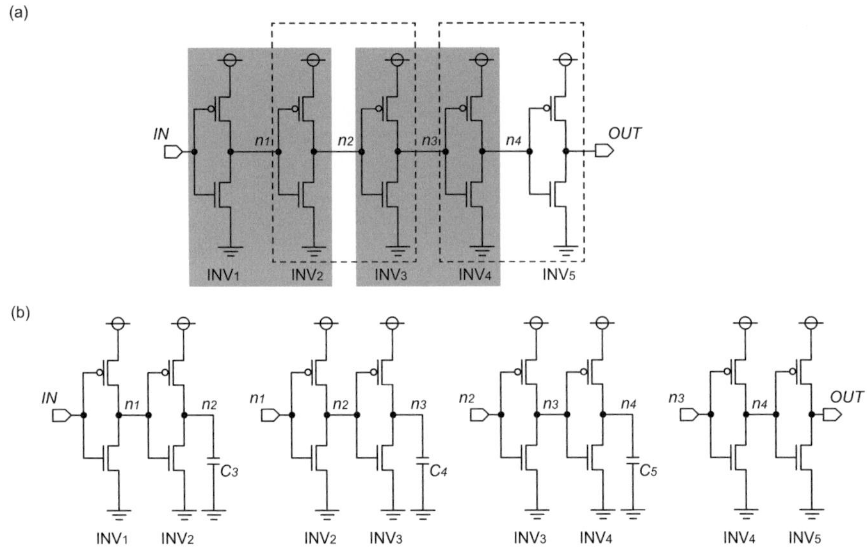

Fig. 2.13 Partitioning with increased accuracy

As for the simulation time, it is faster to simulate a single stage inverter circuit five times than it is to simulate a five-stage inverter circuit simultaneously. As we can tell from Eq. (2.5), the size of the matrix in the generated circuit equation is proportional to the square of the number of nodes. Therefore, if the number of elements reduces to 1/5, the size of the matrix becomes 1/25, and while it does depend on the calculation method, the computational complexity of the matrix inversion becomes 1/125. Even if this is repeated five times, the total computational complexity is 1/25, which means we can expect an acceleration of 25 times.

In cases where the majority of the circuitry is at sleep and only a portion of the circuit is operating, such as in the readout of a memory circuit, event-driven simulation is used for a few of the partitions, drastically improving the simulation speed.

A partitioning with increased accuracy is indicated in Fig. 2.13. In this example, to calculate the change in the first stage n_1, the behavior of INV_3 is ignored, but the state of INV_2 is fully considered. The result calculated for n_2 here is thrown away, and the more accurately calculated waveform for n_1 is used as the input to the next stage consisting of INV_2 and INV_3 to calculate n_2. We can repeat this process of considering inverters two stages away to be constant capacitances, but taking into account the bias conditions of the inverter one stage afterwards. This will increase the computation time, compared to how calculations occurred in Fig. 2.12, but the simulation is more accurate. In normal Fast SPICE, the partitioning size is changed based on the specified accuracy.

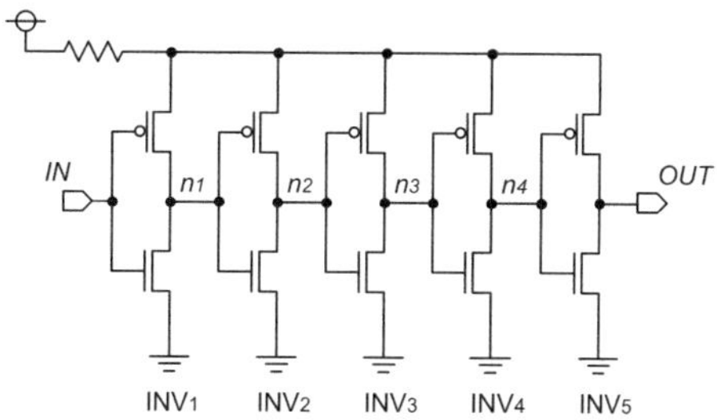

Fig. 2.14 An unpartitionable circuit example

2.2.2 *Unpartitionable Circuits*

We understand that partitioning can accelerate simulations, but there exist certain circuits that cannot be partitioned well. For example, in Fig. 2.12, we could partition the circuit because the behavior of INV_1 and INV_4 are instantaneously unrelated. What about the circuit in Fig. 2.14? In this case, if, for example, INV_4 draws current, the supply voltage of the entire circuit fluctuates and affects the operation of INV_1. Therefore, this circuit cannot be partitioned. It is possible to forcefully partition the circuit by first assuming that the supply voltage remains constant. Then, the voltage and current waveforms can be calculated, and from this calculated current, the supply voltage change can be determined. This option, however, does not take into consideration the effect of supply voltage drop in the delay time calculations, so the simulation is not accurate. Also, depending on the impedance, the supply voltage could be calculated as negative, an obviously unrealistic value. It is important to understand these limitations of Fast SPICE when using these simulation techniques.

2.2.3 *Time Step Control*

There are techniques to determine the time step size in transient simulations. The time step can be lengthened if voltages don't change at all or change very slowly, and shortened if voltage changes appear rapidly. Usually, simulators somewhat automatically control the time step sizes. In simulators with partitioning and event-driven simulations, generally only the partitions (circuit blocks) with activity are simulated. However, if there are partitions with rapid transitions as well as partitions with slow changes, the overall amount of computation is reduced by partition-wise control of the time step, such as increasing the time step for slow-changing partitions (Fig. 2.15a). Without partitioning, the time step needs to be shortened if there is even a single section with rapid changes, so the time step control would be as in Fig. 2.15b. From the comparison of these cases, it is intuitively easy to understand that time step control is more effective with partitioning.

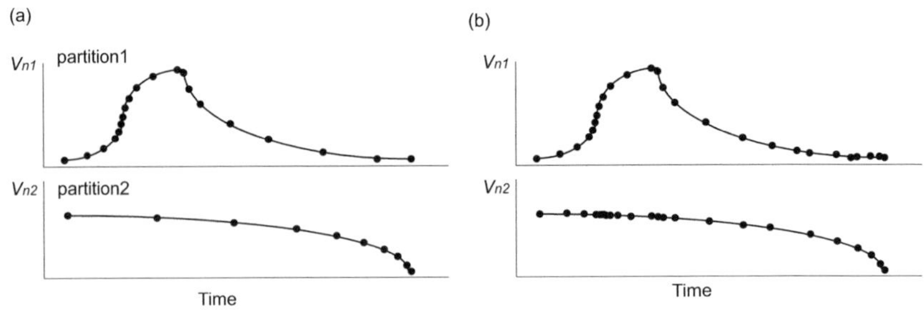

Fig. 2.15 Partition-wise time step control

2.2.4 *Simplification of the Model*

In the BSIM4 model, for example, the analytical equations are extremely complex, and a large amount of time is necessary to calculate the values of the currents. Simulators will save calculation time for the current by simplifying the analytical equations while more or less sacrificing the accuracy. Also, I–V tables for each transistor can be generated ahead of time, and instead of calculating the currents from the voltages through analytical equations, the answer can be looked up in the table to save calculation time. However, this table look-up method may not be used because it can lead to an enormous table size.

2.2.5 *Automatic Decision and Specification of Simulation Accuracy*

Often times logic circuits do not require high precision, but analog circuits do. Fast SPICE can automatically decide whether a circuit is logic or analog based on the netlist and determine the appropriate accuracy (of course, it is possible for the circuit designer to overwrite this and specify the simulation accuracy for each block). Also, the time step interval as a degree of model simplification, as well as the partition granularity (the finer the partitions are, the faster but less accurate the simulations become), can also be automatically determined to an extent.

In regular SPICE, the entire circuit is constantly simulated (Fig. 2.16a), and the behavior of one portion of the circuit can become the bottleneck in terms of the time step for the entire circuit. Therefore, solving a large matrix equation, with size equivalent to the scale of the entire circuit, is required. On the other hand, Fast SPICE only simulates the operating portions of the circuitry, and the time step is also optimally controlled depending on the needs, to lead to the evaluation of small matrix equations. The combination of these techniques make accelerated simulations possible.

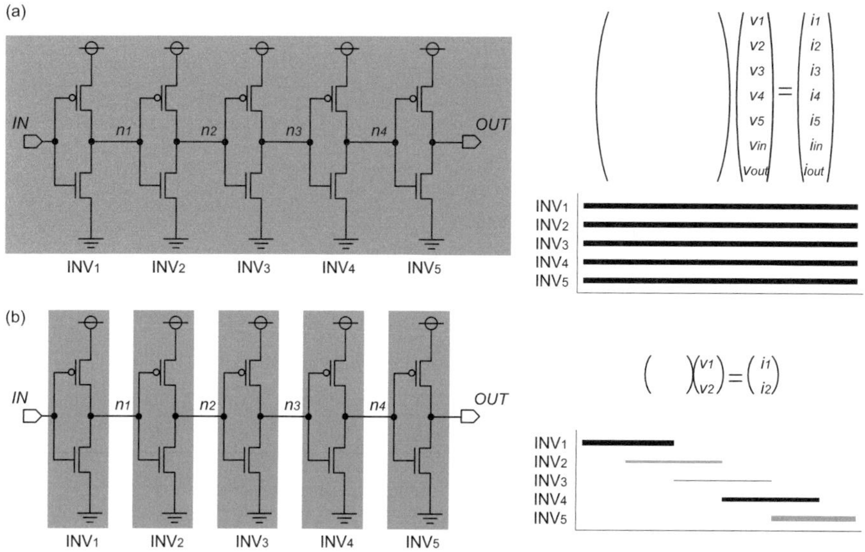

Fig. 2.16 (**a**) Regular SPICE and (**b**) fast SPICE

2.3 A Simple HSPICE Manual

In this section, we take HSPICE of Synopsys as an example to summarize the "netlist" and basic commands used to run simulations. The application to NanoSim is similar.

With regard to the input files, the circuit itself, which will become the LSI, and the files, which indicate the simulation conditions, such as input voltages, temperature, and analysis methods as well as the various options, should be treated as completely separate files.

2.3.1 Basic Points

- All elements and all nodes inside a netlist are given names.
- Node `0` is always the ground potential.
- Either one element or one command is written per line. Inserting + at the beginning of a line will concatenate the line.
- Anything after * or $ in a line will become comments.
- Capital and lowercase letters are not distinguished from one another (`N1` and `n1` are the same node).
- The first letter indicates the element type.
- Lines that begin with `.` (dot) are commands.
- The `.param` command is used to parameterize values.

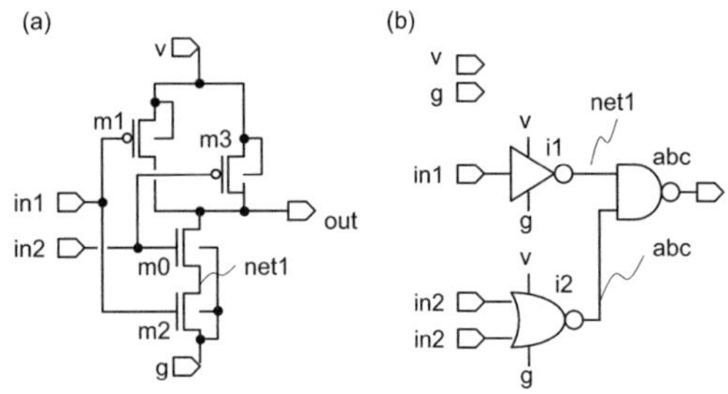

Fig. 2.17 Circuit examples

2.3.2 *Defining Elements*

First, a netlist with the circuit information (e.g., the transistor L and W and the connection information) must be created. Normally the circuit editor will create the netlist for the circuit portion. However, this knowledge can become useful in certain cases, and moreover, any circuit designer should know how to do this.

2.3.2.1 Transistors

M*instName D G S B modelName* L=*length* W=*width* [M=*multi* AD=*drainArea* ...]

Any line that begins with "M" is interpreted as a MOS transistor. From the second character, an independent name is given. The MOS transistor has four terminals, and the nodes are given names in the order: D, G, S, B (the order of D and S can be reversed). After this, the model name, $L =$, $W =$, $M =$, $AD =$, $AS = \ldots$ and so on follows. The model name, for example, would be `NLP` in Sect. 1.2.3. The netlist for the example given in Fig. 2.17a would be as given as follows. The order of elements (switching the `m0` row and the `m1` row) does not affect the results.

```
m0 out in2 net1 g NLP L=180e-9 W=2e-6 M=2
m2 net1 in1 g g NLP L=180e-9 W=2e-6 M=2
m1 out in1 v v PLP L=180e-9 W=5e-6
m3 out in2 v v PLP L=180e-9 W=5e-6
```

2.3.2.2 Subcircuits

.SUBCKT *subcktName nodeName1 nodeName2* ...

...

.ENDS *subcktName*

First, the `.SUBCKT` keyword is used to describe a circuit block as a subcircuit. To describe the NAND circuit of Fig. 2.17a as a subcircuit:

```
.subckt NAND2 g in1 in2 out v
m0 out in2 net1 g NLP L=180e-9 W=2e-6 M=2
```

```
m2 net1 in1 g g NLP L=180e-9 W=2e-6 M=2
m1 out in1 v v PLP L=180e-9 W=5e-6
m3 out in2 v v PLP L=180e-9 W=5e-6
.ends NAND2
```

A subcircuit called `NAND2`, with five terminals `g, in1, in2, out, v` is defined.

2.3.2.3 Instantiating Subcircuits

`X`*instName nodeName1 nodeName2 ... subcktName*
Lines that begin with "X" are interpreted as subcircuit blocks. Node names are referenced in the order used in the `.SUBCKT` definition.

For example, if there were a `NOR2` subcircuit with five terminals `g, in1, in2, out, v` as well as a `INV` subcircuit with four terminals `g, in, out, v`, Fig. 2.17b can be described as follows:

```
Xi1 g in1 net1 v INV
Xabc g net1 abc out v NAND2
Xi2 g in2 in3 abc v NOR2
```

Instance names and node names are allowed to be the same because they can be differentiated in the netlist.

2.3.2.4 Resistors

`R`*instName nodeName1 nodeName2 rVal* [modelName]
Lines that begin with "`R`" are interpreted as resistors. To use a resistor model whose resistance value depends on the temperature or current, the model name is specified in *modelName*, but if this is omitted a simple resistor with the relationship of $V = IR$ is used.

2.3.2.5 Capacitors

`C`*instName nodeName1 nodeName2 cVal* [modelName]
Lines that begin with "`C`" are interpreted as capacitors. To use a capacitor model whose capacitance value depends on the temperature or voltage, the model name is specified in *modelName*, but if this is omitted, a simple capacitor with the relationship of $V = \frac{1}{C}\int Idt$ is used.

2.3.2.6 Inductors

L*instName nodeName1 nodeName2 lVal* [modelName]

Lines that begin with "L" are interpreted as inductors. To use an inductor model whose inductance value depends on the temperature or voltage, the model name is specified in *modelName*, but if this is omitted, a simple inductor with the relationship of $V = L\frac{dI}{dt}$ is used.

2.3.3 Voltage and Current Sources

Sometimes voltage and current sources are placed as symbols in the circuit schematic, but these signals are given from the outside in a real situation. Therefore, these should be treated as simulation conditions and be separated from the circuit schematic. Aside from the circuit information that is output from the circuit schematic editor, what we deal with from here on is necessary knowledge to specify the simulation conditions by, for example, directly editing text files.

2.3.3.1 Voltage Sources

V*instName nodeName1 nodeName2* [etc...]

Lines that begin with "V" are treated as voltage sources. There are many types of voltage sources and can be set to have varying functionalities in DC, AC, and TRAN simulations.

```
.param mvdd = 1.8
.param v1 = 0
.param v2 = mvdd
.param tdelay = 1n
.param tr = 100p
.param tf =  90p
.param freq = 100MEG
.param period = '1/freq'
.param voffset = 'mvdd/2'
.param vamp = 0.1
.param time1 = 0
.param time2 = 1n
.param time3 = 1.5n
.param volt1 = 0.2
.param volt2 = 0.8
.param volt3 = 1.0

Vgnd gnd 0 DC 0
Vdd vdd gnd DC mvdd
Vdd1 vdd1 gnd DC 2.0
Vinp inp gnd DC 'mvdd/2' AC 1 0
Vinn inn gnd DC 'mvdd/2' AC 1 180
```

```
Vin1 in1 gnd DC 1.8 AC 1 PULSE(v1 v2 tdelay tr tf pw
period)
Vin2 in2 gnd SIN(voffset vamp freq tdelay)
Vin3 in3 gnd DC mvdd PWL(time1 volt1, time2 volt2, time3
volt3)
```

In HSPICE, nodes with the name `0` are absolute ground. In the line that begins `Vgnd ...`, the node `gnd` and the node `0` are shorted (connected by 0 V).

In this example, variables are defined and used for most values, using `.param`. It is convenient to do so when running simulations, but it is also possible to directly write in values, as in the line that begins with `Vdd1 ....`

`Vdd` is a 1.8 V source. `Vinp, Vinn` are 0.9V voltage sources in DC and transient analyses, and complementary inputs of amplitude 1 V in AC analysis. `Vin1` is a constant voltage source of 1.8 V in DC analysis, a sinusoidal voltage source of bias 1.8 V, amplitude 1 V, and zero phase in AC analysis, and a trapezoidal waveform between 0 and 1.8 V, as indicated in Fig. 2.18a, in TRAN analysis. `Vin2` is a sinusoidal wave centered around 0.9 V with an amplitude of 0.1 V, as indicated in Fig. 2.18b, in TRAN analysis. `Vin3` is a voltage source with a waveform in which the points `(0s, 0.2V), (1ns, 0.8V), (1.5n, 1.0V)` are connected by straight lines, as shown in Fig. 2.18c, in TRAN analysis. There is no limit to the number of points that can be defined in PWL.

There are many other types of voltage sources, so look these up in the simulator manual as necessary, but in most cases, the previous examples should suffice.

2.3.3.2 Current Sources

`I`*instName nodeName1 nodeName2* [etc...]

Lines that begin with "`I`" are interpreted as current sources. The basics are the same as voltage sources, and DC, AC, SIN, PULSE, and PWL, among others, can be used.

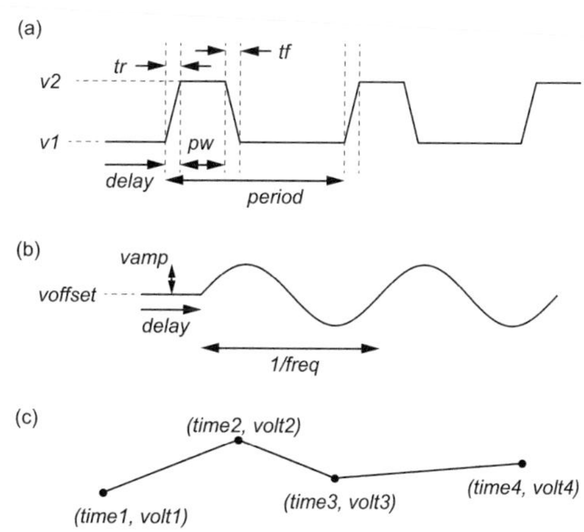

Fig. 2.18 Voltage source waveforms (**a**) PULSE, (**b**) sin, (**c**) PWL

2.3.4 Simulation Types

2.3.4.1 DC Analysis

`.DC` `V`*instName fromVol1 toVol1 stepVol1 VinstName2 fromVol2 toVol2 stepVol2* ..

The output waveform file of HSPICE has the filename `***.dc0`.

It is common in DC analysis to sweep the input voltage and calculate the change in the output voltage. How to sweep which voltages are specified using the `.DC` command. The voltage source `V`*instName1* is swept from *fromVolt* to *toVolt* in steps of *stepVol*, and DC analysis is repeatedly conducted. Also, double, or triple, sweeps are possible. To obtain multiple $I_D - V_D$ curves for various V_D at 0.01 V steps and by changing V_G by 0.1 V steps:

```
.DC VD 0 mvdd 0.01  VG 0 mvdd 0.1
```

2.3.4.2 AC Analysis

`.AC [DEC|LIN]` *numOfPoints fromFreq toFreq*

The output waveform file of HSPICE has the filename `***.ac0`.

The input signals to AC analysis are included in the voltage source descriptions `V***`, so the frequency at which to conduct the AC analysis is specified in the `.AC` command.

```
.AC DEC 10 1K 10G
```

In the above line, the frequency is swept from 1 K[Hz] to 10 G[Hz] in equal step sizes on the log scale, and 10 points of frequencies are calculated for each decade.

2.3.4.3 Transient Analysis

`.TRAN` *timeStep endTime*

The output waveform file of HSPICE has the filename `***.tr0`.

In transient analysis, the basic time step and the end time are specified. As a rule of thumb, the time step should be set to about 1/100∼1/1000 of the CLK period. The time step specified here is only a guide, and the simulator will control the fine details accordingly.

2.3.4.4 Harmonic Balance Analysis

`.HB TONES=`*freq* `NHARMS=`*numOfHarms*

Harmonic balance analysis cannot be conducted with HSPICE; HSPICE-RF is necessary. The output waveform file has the filename `***.hb0`.

The input base frequency (TONES) and the number of higher-order harmonics to consider (NHARMS) are specified.

```
.HB TONES= 100MEG NHARMS= 8
```

In the above line, to an input of 100MEG, up to the eighth-order harmonics (100MEG, 200MEG, ..., 800MEG) due to the circuit nonlinearities are calculated.

2.3.5 *File Includes and Libraries*

`.include "`*dirName/fileName*`"`
`.lib "`*dirName/libraryFileName*`"` *libName*

A simulation netlist is often created as several divided files which are joined together to be interpreted as a single netlist. The `.include` command is used to include the specified file. HSPICE will by default automatically convert all letters within a file to lowercase, but case sensitivity is maintained in sections of the `.include` line between double quotation marks.

Normally, SPICE parameter files are distributed as separate files from SPICE simulation control files and netlist files, and these SPICE parameter files are used by including them. Here, as explained in Sect. 1.2.3, parameters can be defined in libraries of separate files for each process corner. In this case, after including the SPICE parameter file, the file with the library definition as well as the library name are specified by the `.lib` command. For example, if as in Sect. 1.2.3 the library name is NT:

```
.include "../rules/vdec1.par"
.lib "../rules/vdec1.lib" NT
.lib "../rules/vdec1.lib" PT
```

The SPICE parameter file `../rules/vdec1.par` is included, and the NT and PT libraries (process corner conditions) defined in the file `../rules/vdec1.lib` are used in simulation.

There are cases where the process corner libraries are included within the SPICE parameter file. In this case, the `.lib` line is sufficient and `.include` is unnecessary.

2.3.6 *Options and the* `.MEASURE` *Command*

2.3.6.1 Options

There are various options besides the basic commands. Some common ones are listed below.

`.OPTION POST <=2>` A waveform file is output. If =2, the output waveform file is output not in

	binary but rather in ASCII text format (the default 1 is binary output).
`.OPTION POST_VERSION=2001`	The significant digits of the output waveform file is increased from 5 to 7 digits. This is used often when the simulation time step is extremely small and the significant digits of time needs to be increased.
`.OPTION PROBE`	The nodes for which the waveform will be output is limited to the nodes specified by `.PROBE`. In HSPICE the option `.OPTION POST` will output all nodes, but in Nanosim using this option along with `.PROBE` as in the example below is necessary.
`.PROBE V(`*nodeName*`) I(`*instName*`)`	When the option `.OPTION PROBE` is specified, the nodes for which the output should be saved to the waveform file is listed. In Nanosim this option must be used with `.OPTION PROBE` as a set. Wildcard characters `*` can be used for *nodeName* and *instName*, for example, as in `.probe V(out*) V(Xinst1.int*) I(V*)`.
`.OPTION POSTTOP` <=*N*>	The node waveforms from the top of the SUBCKT layer to the *N*th layer are output. If `POSTTOP=1`, only the nodes of the upper most layer are saved. However, this cannot be used in conjunction with `.OPTION PROBE`. If both are specified, `.OPTION PROBE` has priority.
`.OPTION POSTLVL` <=*N*>	The node waveforms of the *N*th layer only are output. This also cannot be used with `.OPTION PROBE`. If both are specified, `.OPTION PROBE` has priority.
`.OPTION ACCURATE`	More accurate simulation values are obtained by making the convergence detection conditions stricter. The time it takes to simulate will become longer.
`.OPTION RUNLVL=`*N*	Simulation times and accuracies are controlled by controlling the strict-

	ness of convergence detection conditions. N takes on values between 1 and 6, where 1 is the fastest and 6 is the most accurate. The default value is 3.
`.OPTION MTTHRESH=`*number*	If the simulation is specified to use multiple CPUs from the command line but the number of elements is fewer than what is specified in this option, then the simulation only uses 1 CPU. The default value is 16.
Options for Analog Artist	To cross-probe the waveforms simulated in HSPICE with Cadence Analog Artist: `.OPTION INGOLD=2 ARTIST=2 PSF=2 HIER_DELIM=1`

2.3.6.2 The .MEASURE Command

There are cases during simulation when we would like to find out, for example, the oscillation frequency of a ring oscillator or the delay from input to output. We may also want to know, when designing an opamp, the unity gain frequency, the phase margin, and the bandwidth. It is common practice to observe this on the waveform viewer by using markers, but by giving appropriate commands beforehand, the simulator itself will calculate and return values in simulation. The .MEASURE commands that I frequently use are listed below.

```
.MEASURE TRAN charge INTEG I(VOUT)
.MEASURE TRAN upwidth TRIG V(UP) VAL='mvdd/2'
+                  RISE=1 TARG V(UP) VAL='mvdd/2' FALL=1
.MEASURE TRAN dnwidth TRIG V(DN) VAL='mvdd/2'
+                  RISE=1 TARG V(DN) VAL='mvdd/2' FALL=1
.MEASURE TRAN level FIND V(n5) AT=5m
.MEASURE TRAN _tranVdiffF PP V(OUTTRAN) FROM='td+period'
+                  TO='td+period*2+td'
.MEASURE TRAN _tranVminF MIN V(OUTTRAN) FROM='td+period'
+                  TO='td+period*2+td'
.MEASURE TRAN _tranFstart WHEN
+                 V(OUTTRAN)='_tranVminF+0.9*_tranVdiffF'
+                 TD='td+period' FALL=1
.MEASURE TRAN _tranFend WHEN
+                 V(OUTTRAN)='_tranVminF+0.1*_tranVdiffF'
+                 TD='td+period' FALL=1
.MEASURE TRAN slewF
+         PARAM='_tranVdiffF*0.8/(_tranFend-_tranFstart)'

.MEASURE TRAN _tranDelayR TRIG V(OUTTRAN)
+                 VAL='_tranVminR+0.1*_tranVdiffR'
+                 TD='td+period' RISE=1 TARG V(OUTTRAN)
```

```
+                VAL='_tranVminR+0.9*_tranVdiffR' RISE=1
.MEASURE TRAN slewR PARAM='_tranVdiffR*0.8/_tranDelayR'

.MEASURE DC outVoltRange PP par('V(OUTDCAC)')
+                FROM='0.0' TO='mvdd'
.MEASURE DC idd  FIND par('-I(VV)') AT='mvdd/2'
.MEASURE DC offset  FIND V(INDCAC, OUTDCAC) AT='mvdd/2'

.MEASURE AC gain FIND Vdb(OUTDCAC) AT=1k
.MEASURE AC phaseMargin FIND par('180+Vp(OUTDCAC)')
+               WHEN Vdb(OUTDCAC)=0
.MEASURE AC unitGainFreq WHEN Vdb(OUTDCAC)=0
.MEASURE AC cmrr FIND par('Vm(OUTCMRR)/Vm(OUTDCAC)') AT=1k
.MEASURE AC psrr FIND par('Vm(OUTPSRR)/Vm(OUTDCAC)') AT=1k
```

2.3.6.3 Other Common Commands

`.IC V(`*nodeName*`)=`*volt* — An initial condition (the value at $t = 0$ in a TRAN simulation) for a node voltage is specified. In a ring oscillator simulation, this must be specified at some node or else all nodes will be at half V_{DD} in a metastable state and will not oscillate.

`.PARAM` *varName* `=` *val* — Values such as the input voltage or frequency of a voltage source, or a transistor's L and W, are specified as parameters. For example,

```
.param mvdd = 1.8
.param trtf = 100p
.param freq = 100MEG
.param period = '1/freq'
Vin in gnd PULSE(0 mvdd 0 trtf trtf
+        'period/2-trtf' period)
```

To compute a parameter, use single quotation marks `'`.

`.vec "`*fileName*`"` — The input waveform shape can be given as a vector file of 0's and 1's. If the line `.vec "aaa.vec"` is added to a regular input file, and the file `aaa.vec` is written as

```
RADIX 1111
IO    IIII
VNAME BIN_2 BIN_1 INPUT CTRL
TUNIT ns

VIH 1.2
VIL 0
SLOPE  0.01
PERIOD 1
```

```
0011
1011
1000
0011
1000
1011
```

then terminals connected to voltage sources `BIN_2, BIN_1, INPUT, CTRL` are defined, and an input waveform that transitions between 1.2 and 0 V with a rise time and fall time of 10 ps and 1 ns, respectively, is created.

2.3.6.4 Command Line Options

`-i` *`fileName`*	Specify the input file name.
`-o` *`dirName`*	Specify the directory for the output file.
`-mt` *`N`*	Specify the number of cores to use in a multi-core simulation. However, this comes at the cost of occupying more licenses.
`-hpp` (2010.03.SP-2 and later)	Accelerated algorithms are used. This is also optimized for multi-core, and the computational efficiency does not saturate with increased number of cores. However, this occupies an excess number of licenses.

The machine I regularly use is a Quad Core, so I often use:

```
% hspice -hpp -mt 4 -i fileName.sp
```

as a standard command.

Chapter 3
Layout and Verification

Once the circuit design is finished, we begin the layout design. You're given a thick layout manual and casually told, "Next up for you is layout." A large amount of creativity is not necessary, but rather a persistent and tenacious effort is required, and within the design cycle, this step is "painful" and takes the most time. In turn, it is no exaggeration to say that this step determines whether the created chip will function successfully or not.

3.1 The Basic Process of LSI Fabrication

In proceeding with layout, it is first necessary to know the three-dimensional structure of LSI and the corresponding fabrication process. The fabrication process can be divided into photolithography, deposition, doping, and a removal process.

3.1.1 The Three-Dimensional LSI Structure

Figure 3.1a indicates the three-dimensional structure of LSI, and Fig. 3.1b shows the cross section. A polysilicon gate is made on top of the thin gate oxide layer of the silicon surface. Ions are implanted for the source and drain regions, and the source, drain, and gate regions are processed with silicide to reduce the resistance. Structures such as contact holes for transistor terminal connections, metal wires, and via holes for vertical wire connections are created for the final LSI circuit structure.

T. Nakura, *Essential Knowledge for Transistor-Level LSI Circuit Design*,
DOI 10.1007/978-981-10-0424-7_3

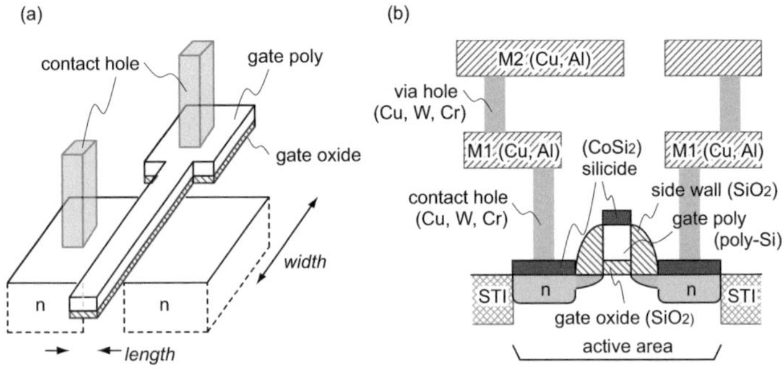

Fig. 3.1 The transistor structure

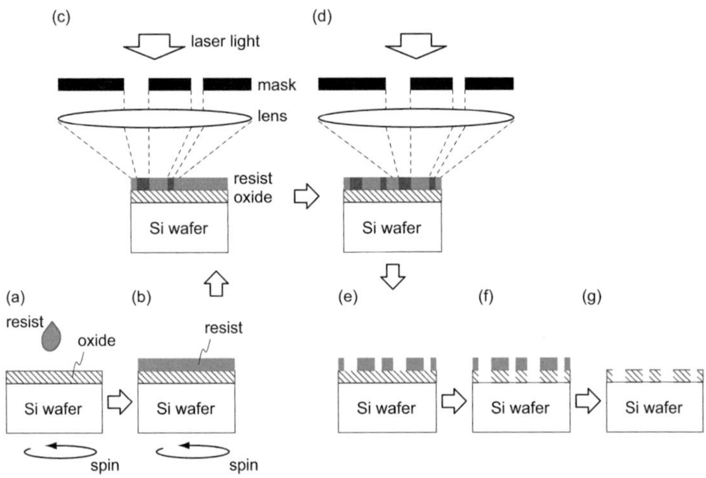

Fig. 3.2 Manufacturing process. (**a**) (**b**) Applying photoresist, (**c**) (**d**) photolithography, (**e**) development, (**f**) etching, (**g**) resist removal

3.1.2 *Photolithography*

3.1.2.1 Basic Principle

In the process of manufacturing LSI, the three-dimensional structure is built from the bottom-up as shown in Fig. 3.1. For example, the process of creating a partial oxide film is shown in Fig. 3.2. (a) Liquid resist is dropped onto the silicon wafer while it is spun at a fast speed. (b) Centrifugal force is used to create a thin film of resist on top of the wafer. (c) Laser light is irradiated onto a mask with the design pattern drawn, and by collecting the light onto the wafer with a lens, a fine design pattern, which is the mask pattern shrunk, is irradiated. (d) This pattern is repeatedly drawn on the wafer surface by moving the wafer. (e) Since the resist characteristics only change where light has hit, only those portions (or the portions where light did not hit) are selectively removed. (f) Oxide film regions that are not protected by resist are selectively etched away. (g) The resist is removed.

With photolitography techniques, the process in Fig. 3.2 is repeated for selective chemical processing, which allows the necessary materials to be created at the necessary places.

There are two types of resist. The type that is used to remove only the portions hit with light and keep the portions not hit by light is called positive resist, and the type that is used to keep the portions hit by light is called negative resist.

3.1.2.2 Wavelength of the Light Source

The mask pattern is transferred to the resist on the wafer by laser light, as in Fig. 3.2c. Due to the wave nature of light, a pattern finer than the wavelength of the laser light used as the light source cannot be transferred cleanly. Therefore, the limit to fine patterns is determined by the laser light wavelength.

3.1.2.3 Phase Shift Mask

To transfer patterns with a finer pitch than the light source, phase shift masks are used. Normally the pattern is transferred as in Fig. 3.3a, and the resist receives a light intensity distribution whose skirts are spread as much as the wavelength. By using a mask which shifts the phase by half a wavelength as in Fig. 3.3b, the electric fields from adjacent patterns cancel out, enabling the clean transfer of patterns. However, phase shift masks are more expensive than ordinary masks, so these are often only used for specific masks such as for gates and contacts.

3.1.2.4 Mask Alignment Errors

In the manufacturing process, the layered structure is built from the bottom-up. However, there can be errors when the mask is aligned, and the pattern to be

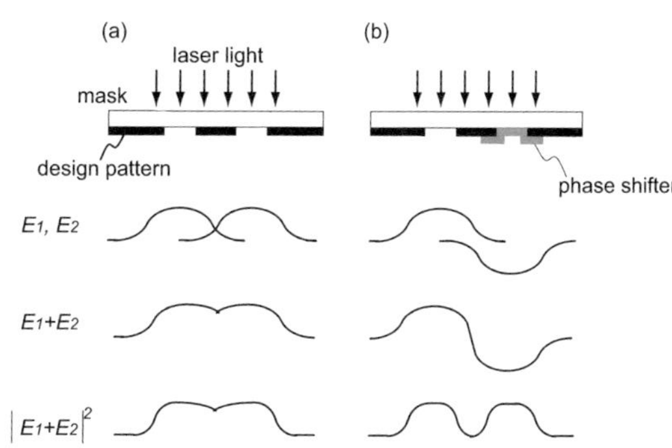

Fig. 3.3 Phase shift mask

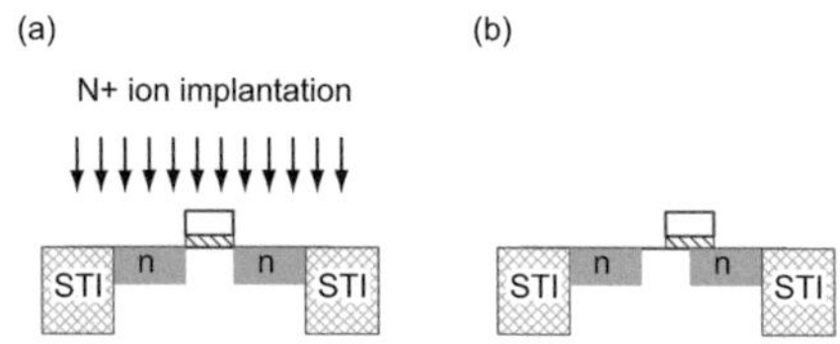

Fig. 3.4 (**a**) Self-alignment, (**b**) alignment error

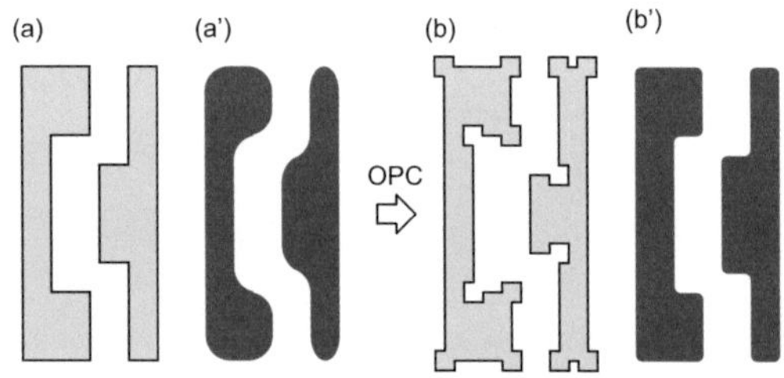

Fig. 3.5 (**a**) Layout design, (**a′**) final design without OPC, (**b**) layout design with OPC, (**b′**) final design with OPC

transferred can deviate from the targeted position. It is necessary to consider this deviation during layout design, and how much deviation can occur and how much margin is needed to ensure circuit functionality are prescribed in the "design rule manual" (described later) in great detail.

3.1.2.5 Self-Alignment

Also, the manufacturing process is devised so that the mask alignment accuracy can be loosened as much as possible, by taking into consideration the mutual relationships between layers. This is referred to as "self-alignment." For example, as in Fig. 3.4a, N+ ion implantation is done after forming the gate and STI, which naturally aligns the gate edge and source and drain edges, thus preventing the case shown in Fig. 3.4b.

3.1.2.6 OPC

When creating shapes that have dimensions in the tens of nanometers, the manufacturing based on the layout design of Fig. 3.5a will yield the result shown in Fig. 3.5a′ due to the effects of the accuracy of lithography and etching. By compensating from Fig. 3.5a, b, a shape that is closer to the designer's intentions, as in Fig. 3.5b′, can be manufactured. This technique is called optical proximity correction (OPC).

Normally, the layout designer will design the shape of (a), while the correction from (a) to (b) is conducted under the responsibility of the manufacturer. When the process engineers are designing the fabrication process, several typical sample patterns are fabricated and a database containing design shapes and final shapes is created. Based on this, masks are fabricated with the optimal OPC using a specialized CAD tool.

3.1.3 Deposition

There are three main types of deposition or film formation.

3.1.3.1 Thermal Oxidization

When silicon is placed in an oxygen atmosphere with high temperature, the silicon oxidizes and forms SiO_2. This method allows for good control of the oxide thickness and yields high-quality oxide layers, and thus gate oxide layers are manufactured with this method.

3.1.3.2 CVD

Chemical vapor deposition (CVD) deposits compound material on the silicon wafer by the reaction of the silicon surface with a reactive gas flowing on the wafer. Polysilicon for the gate poly, for example, is deposited through the reaction of $SiH_4 \rightarrow Si + 2H_2$.

3.1.3.3 Sputtering/PVD

Electrons are discharged in argon gas at low pressure (close to vacuum), and accelerated electrons are bombarded onto the target aluminum by the Lorentz force. Aluminum atoms that are emitted due to the collision are deposited onto the silicon wafer, as shown in Fig. 3.6. This method is called sputtering, or physical vapor deposition (PVD).

3.1.4 Removal of Unnecessary Parts

The unnecessary parts of formed or deposited layers are removed. There are two types: etching and CMP.

Fig. 3.6 How sputtering works

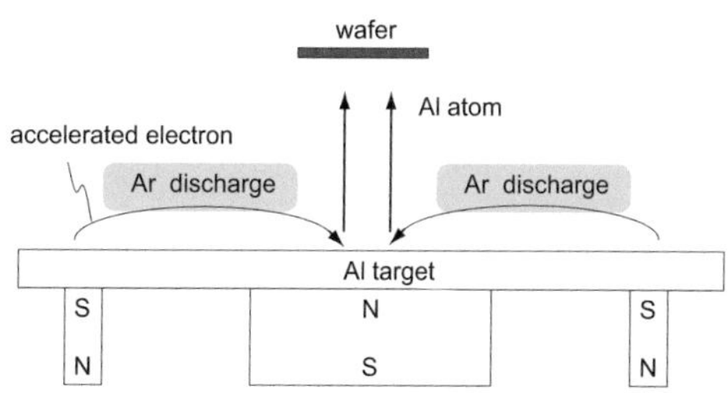

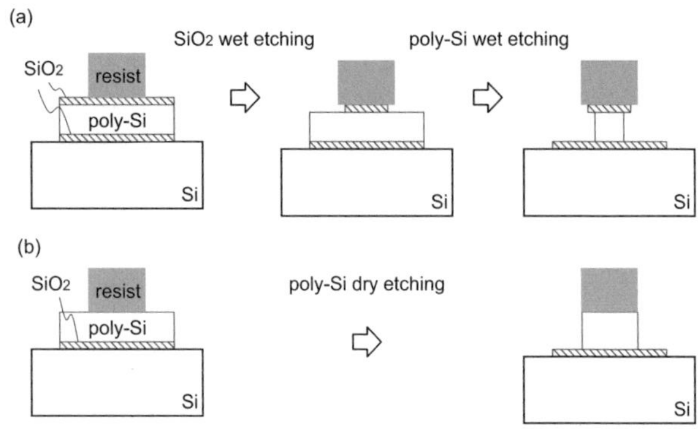

Fig. 3.7 (**a**) Wet etching, (**b**) dry etching

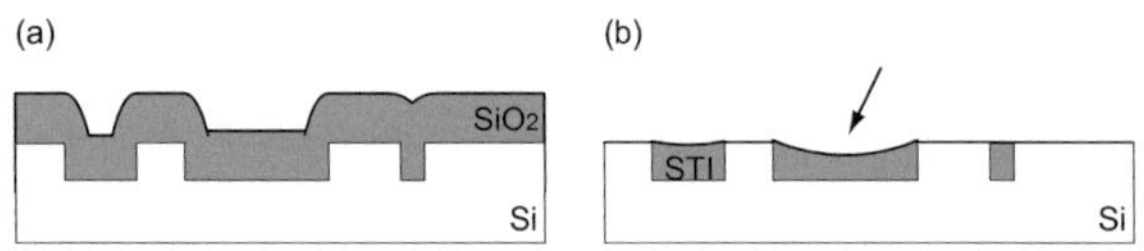

Fig. 3.8 (**a**) Before CMP, (**b**) after CMP

3.1.4.1 Etching

There are two types of etching: wet etching, which is the removal through immersion in a solution, and dry etching, which uses plasma instead, for example, in reactive ion etching (RIE). As indicated in Fig. 3.7, dry etching generally results in sharper cuts.

3.1.4.2 CMP

Chemical mechanical polishing (CMP) flattens the surface by chemically dissolving the surface and a sandpaper-like mechanical polishing. However, as the arrow of Fig. 3.8b shows, polishing a wide surface results in dents and uneven surfaces. Therefore, layout design must be conducted so that polishing does not need to be done over regions larger than a certain area. This is the origin of "density rules," which we will discuss later.

For example with STI, the silicon is selectively etched, and after an oxide layer is deposited with CVD, the surface is flattened with CMP.

3.1.5 *Introduction of Impurities*

3.1.5.1 Ion Implantation

When group III ions such as boron (B) are implanted into the silicon wafer with high energy by means of a high voltage, the wafer becomes an n type. When group V ions such as phosphorus (P) are implanted, the silicon becomes p type.

3.1.5.2 Annealing

Directly after ion implantation, the silicon crystal has cracks due to the implanted ions. However, leaving the wafer in a high-temperature inert gas for several tens of seconds will seal these cracks, as well as allow the implanted ions to spread evenly in a Gaussian distribution. This process is called annealing and is often conducted after ion implantation.

3.1.6 *CMOS Fabrication Process*

The flow of CMOS fabrication process is indicated in Fig. 3.9. Some resist remains on the surface after development from the removal process. Usually a p-type substrate is used, and first the (a) N-well and (b) P-well are formed via ion implantation. Next, (c) STI is formed in the non active regions by silicon etching, CVD oxidization, and CMP. (d) The gate oxide layer and gate poly are deposited through thermal oxidization and CVD, and the gate is created with etching. (e) N^- and (f) P^- ion implantations for the lightly doped drain (LDD) are conducted. (g) The gate side walls are created. Source, drain, and body contact regions are created with (h) N^+ and (i) P^+ ion implantations. (j) Cobalt (Co) is deposited onto the regions outside of the oxidized layer and silicided by thermal processing to remove unreacted portions of the oxide film. (k) Interlayer insulator oxide is deposited, flattened with CMP, and buried in tungsten (W) to create contact holes. (l) The M1 region is etched, aluminum (Al) or copper (Cu) is deposited with sputtering to the entire area, and the metal is flattened with CMP to create the M1 wiring. The same process is repeated for M2, M3, etc.

Drawing the necessary patterns for each step as shown on the right side of Fig. 3.9 is what we call layout design.

3.1.7 *Dual Damascene*

The explanations of Fig. 3.9 are with the assumption that aluminum (Al) will be used for wiring, but in recent years, copper (Cu) wiring is used to reduce the wiring

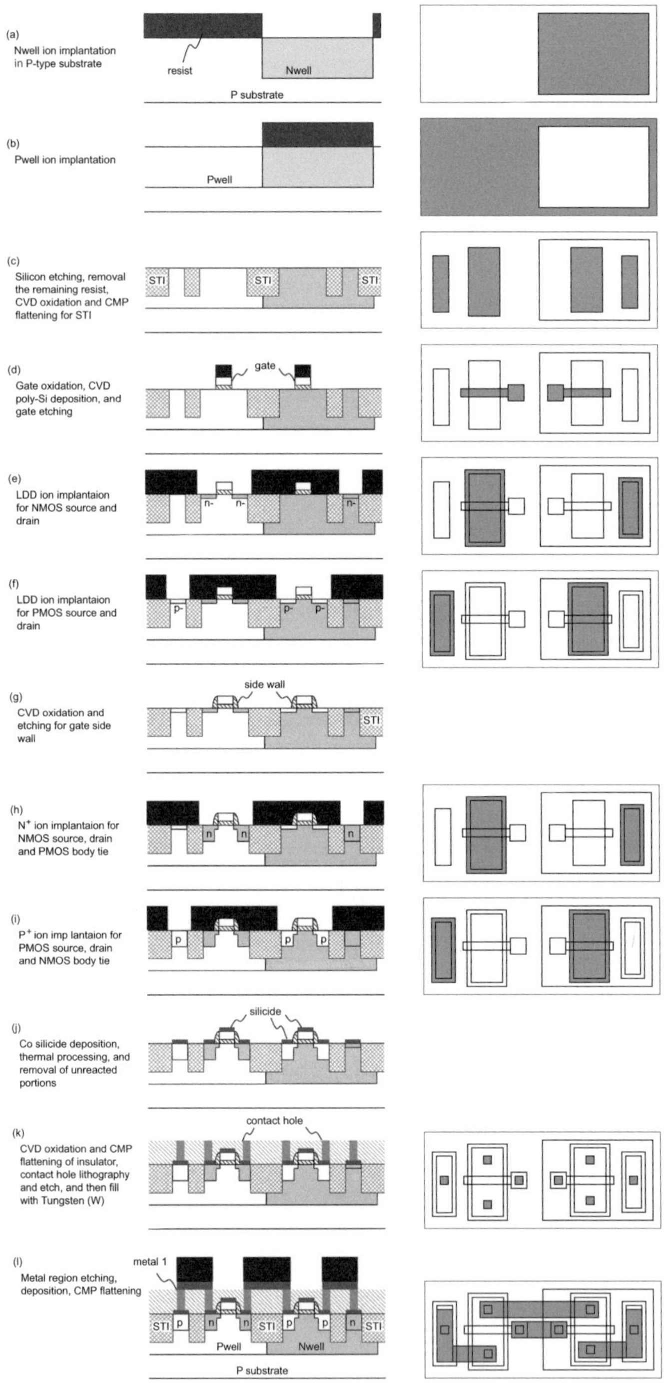

Fig. 3.9 CMOS fabrication process

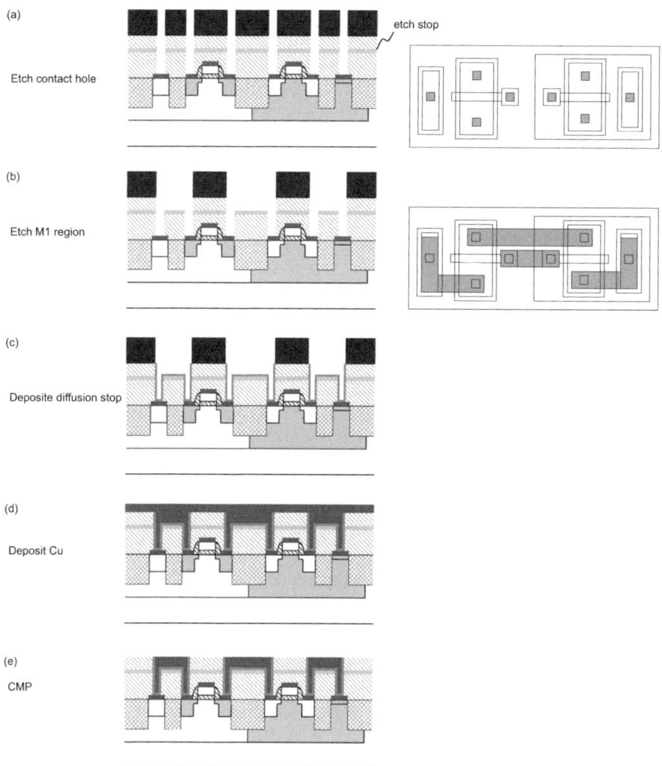

Fig. 3.10 Dual damascene

resistance. A method called dual damascene is used to create copper wiring. This is a method to solve these problems:

- It is difficult to etch copper.
- Copper tends to diffuse into the oxide film.

and is used widely for copper wiring in processes beyond 90 nm.

First, as shown in Fig. 3.10, (a) an interlayer insulator, etch stop layer, and wiring layer insulator are deposited. Then, patterns for the contact holes are developed by photolithography, and contact holes are drilled out. Next, (b) wiring patterns are developed by photolithography, and the wiring area is etched. The etching is stopped at the etch stop layer. (c) A diffusion stop layer is formed to prevent spreading of the copper. (d) Copper for both the contact hole and wiring portions are deposited by sputtering. (e) Excess portions are removed with CMP.

In general, the burying of etched portions and removal of excess with CMP are called the damascene process. The dual damascene process gets its name because two layers, the contact hole and wiring, are buried at the same time.

3.2 Design Rules

Rules which must be followed to ensure proper circuit operation, while considering the uncertainties due to mask alignment errors and limits in the light source wavelength for photolithography, are prescribed in the design rule manual.

3.2.1 *Basic Rules*

3.2.1.1 Types of Basic Rules

There are six types of rules. The top of Fig. 3.11 indicates the data, and the bottom shows the actual shape that is formed.

Minimum width (a) May break if too narrow
Minimum spacing (a)(b) May touch if too close
(This is called min. space for within the same layer and min. clearance for between layers)
Minimum extension (c) May not extend out if too short
Fixed size (b) Contact and via sizes are fixed
Minimum area (d) May disappear if too small
Maximum width The chip may bend if too wide

3.2.1.2 Basic Rule Examples

The rules to be observed for each layer are listed, such as the following: the wire should be wider than 0.4 μm, the gate poly should extend beyond the active area (AA) by more than 0.2 μm, and so on. For example, the page regarding the gate poly may look like that shown in Fig. 3.12 (the values in the figure are not necessarily

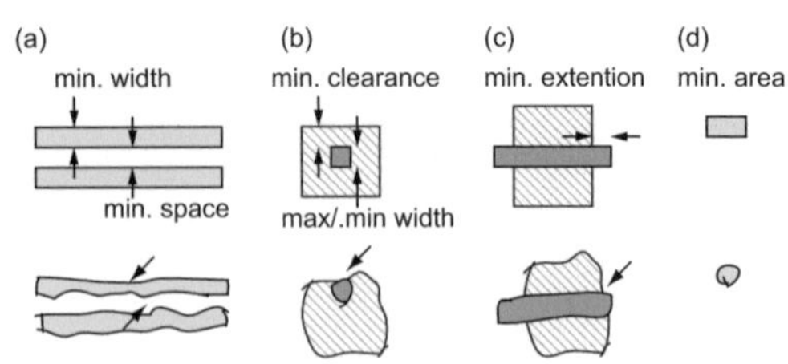

Fig. 3.11 Basic rule types

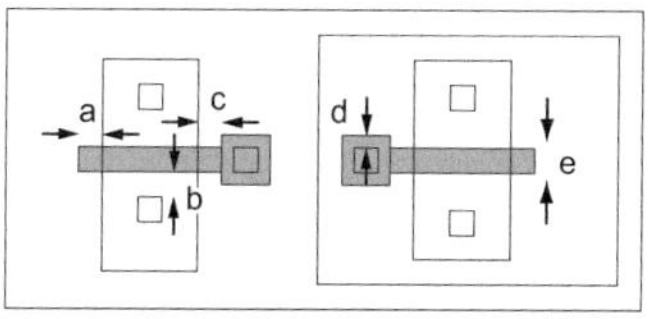

a	Min. extension of GATE beyond AA	0.2um
b	Min. clearance from CH to GATE	0.4um
c	Min. clearance from AA to GATE	0.1um
d	Min. extension of GATE beyond CH	0.1um
e	Min. width of GATE	0.2um

Fig. 3.12 Design rule examples

realistic). In this way, the rules for each layer N-well, active area, N^+ Implant, P^+ Implant, ContactHole, M1, M2, ... VIA12, VIA23, ..., and so on, as well as the relationships between layers, are specified in great detail across dozens of pages.

In recent processes, there are many detailed rules, such as that the M1 area must be larger than 0.02 μm^2 to make sure an island remains after etching or that wirings may not exceed 10 μm to avoid physical stress. Also, there are often times rules whose origins are beyond our imaginations. Therefore, it is important not only to read and understand the manual before starting layout but also to keep the rules on the side during the layout design phase.

3.2.2 The Grid

There is a concept within layout called the grid, and as shown in Fig. 3.13a, all patterns must lie on the grid. You should find out how large the minimum grid is and specify the minimum grid in the layout editor before beginning the layout. It is extremely difficult to fix a shape that has been drawn off grid, as shown in Fig. 3.13b.

The grid may vary depending on the layer, because it is cheaper to create a mask with a coarse grid than it is to create a mask with fine grids. For example, gates and contact holes might be on a fine 0.01 μm grid, whereas the N-well or the top metal layer may be on a coarser 0.05 μm grid.

In addition, it is necessary to check whether 45° diagonal layouts are allowed, whether any arbitrary angles are allowed, and if so how the grid is to be handled in those cases.

3.2.3 Density Rules

In the fabrication process, CMP is used extensively for planarization of the active region or metal interconnect layers. However, as explained in Fig. 3.8, CMP over a

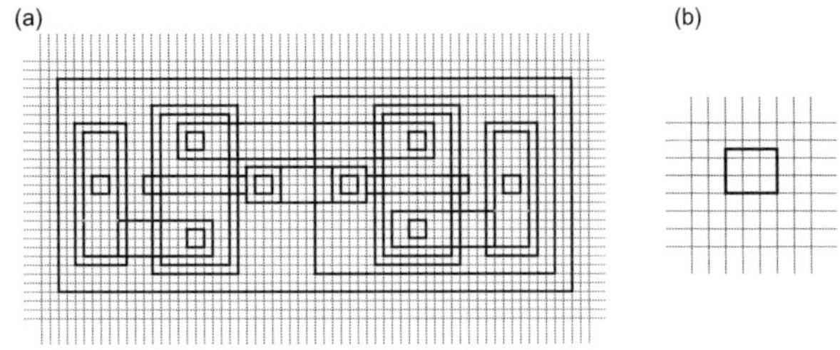

Fig. 3.13 The grid. (**a**) On grid. (**b**) Off grid

Table 3.1 Density rule example

	AA	M1	M2	M3
Entire Chip [%]	40∼60	30∼70	30∼70	20∼80
1 mm Window [%]	30∼70	25∼75	25∼75	20∼80
100 μm Window [%]	20∼80	20∼80	20∼80	20∼80

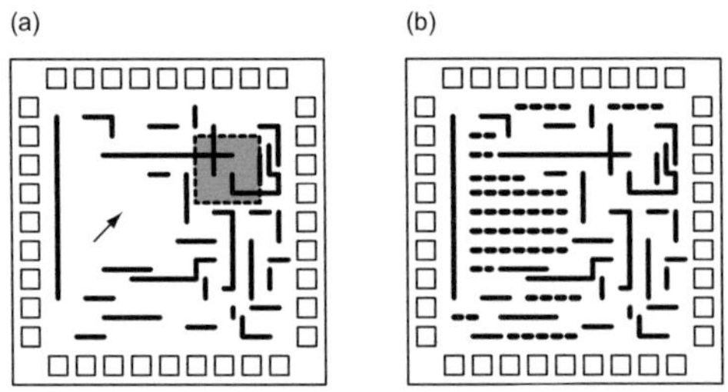

Fig. 3.14 Density rule

wide area results in dents, which results in limits in the density of active regions and wiring layers. These are called the density rules.

There are various levels to the density rules. As indicated in Table 3.1, for instance, the active area density must be 40 %∼60 % over the entire chip, 30 %∼70 % within any arbitrary 1 × 1 mm window, and 20 %∼80 % within any arbitrary 100 × 100 μm window.

For example, if Fig. 3.14a indicates the layout of metal 1, the 1 × 1 mm window indicated by the dotted square can be moved around to calculate the density. We would find that the region indicated by the arrow lacks metal 1 and therefore may introduce dents during CMP which may detrimentally affect the operation of the circuit.

3.2.3.1 Dummy Metal

It is rare to encounter a violation of the density rule on the high side during normal layout design, but it is common to violate the density rule by having a density that is too low. Therefore, a "dummy" wiring is placed that is not connected to anywhere and is not necessary for circuit operation, to meet the density rules for CMP. This is called dummy metal (dummies are also placed in the active region, but there does not seem to be a common name for this). For example, as the dotted dark lines in Fig. 3.14b indicates, dummy metal is placed to fill in the spacings between the original wirings. Contact holes and via holes are not connected, and the state is left as electrically floating.

This dummy metal is not placed by hand by the layout designer. Generally, dummy metal is generated automatically at the last step of layout, and the layout designer does not usually need to worry about this.

However, in cases where the characteristics of a sensitive analog circuit are taken into consideration, such as with spiral inductors (discussed later), there are regions where we would not like to have any dummy metal placed. Normally, along with the dummy metal generation algorithm, a "dummy prohibiting layer" is provided, which allows the designer to specify these regions. In these cases density rules must still be met, and so the designer must place dummy metal by hand to meet density rules within regions where dummy is not automatically generated.

3.2.4 Dummy Transistors

In processes 100 nm and below, there are cases where dummy transistors are necessary. When transistors are arranged in a continuous active region as in Fig. 3.15a, b, the transistor characteristics at the edges may change due to the strain

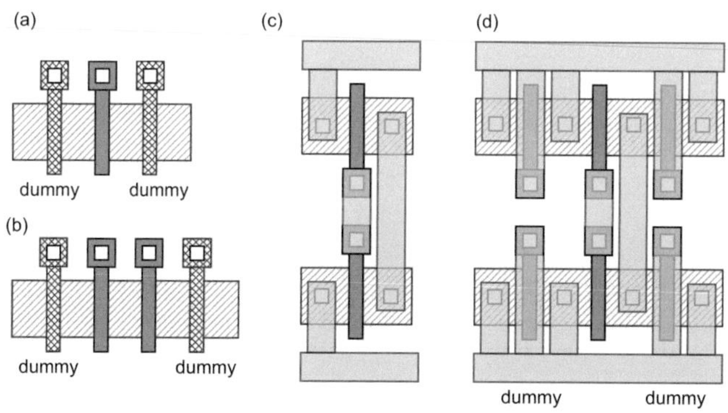

Fig. 3.15 Dummy transistors

from STI or the nonuniformity of etching. Therefore, there are processes with the rule that the edge transistors must be used as dummy transistors and must not be used in circuits. In this cases, the inverter layout would not look like Fig. 3.15c but rather like (d). It is necessary to keep the dummy transistors in the off state (connect the PMOS gate to V_{DD} and NMOS gate to G_{ND}) so as not to affect the other circuitry.

3.2.5 Antenna Rules

Caution is necessary when routing very long wires. Plasma is used in the wiring fabrication process, and charge can accumulate in the wiring during this process. This accumulated charge is discharged through the substrate when a connection to the substrate is made, but if a connection to a MOS gate is made first, then the gate oxide film can be destroyed. This is called plasma-induced gate oxide damage. The antenna rule dictates that the area of wiring that will be connected to a gate before being connected to a substrate must be maintained below a threshold. In the case of Fig. 3.16b, the residual charge on the gray wires will be applied to the gate when M3 is fabricated.

Measures must be taken to fix antenna rule violations, such as raising the wiring on the receiving side to the highest layer as in Fig. 3.16c, limiting the highest metal layer on the sending side as in Fig. 3.16d, or placing an off transistor next to the transistor on the receiving side and connecting the wire to the drain as in Fig. 3.16e.

3.2.6 Electromigration

When a large current flows through a thin wire, the wire can break. The reason is that electrons collide with the metal atoms causing the atoms to shift in position, and this phenomenon is called electromigration. Because this is a shift of atoms due to electron collisions, it is said that the atoms will move back to their original positions if a current is flown in the opposite direction. Therefore, this is not a problem in wiring for logic circuits and clocks that repeat switching between H→L and L→H. Rather, this is problematic for some wires in analog circuitry with constant current flow. For each wiring layer, a maximum current density, such as 1 A/μm, is predetermined. However, these limits are difficult to check using tools (described later), and the circuit designer and layout designer must be cautious and take measures such as to make the routing wider as necessary.

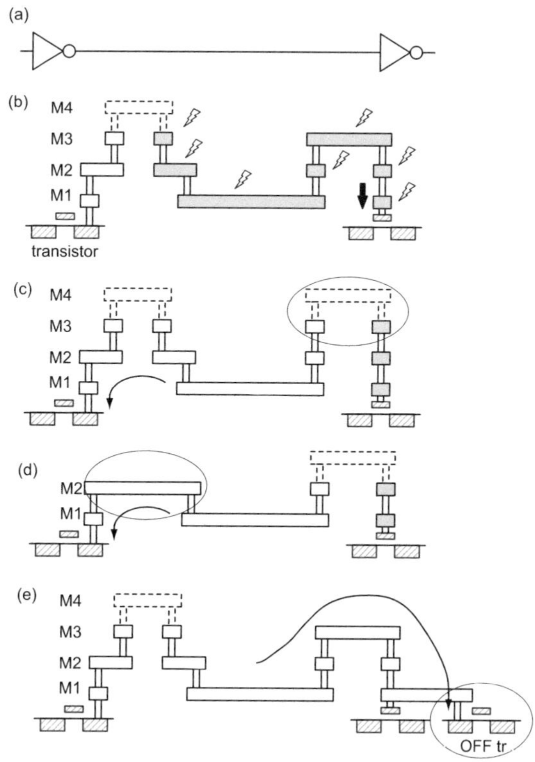

Fig. 3.16 Antenna rules

3.2.7 *Hand-Drawn Layers and Autogenerated Layers*

During layout design, is it really necessary to draw the layout for both P-well and N-well? We could save a lot of trouble in layout by simply drawing the N-well and specifying the P-well to be everywhere except the N-well. Also, the N^+ ion implantation region can be defined, as shown in Fig. 3.17, to be "the active area that a gate crosses over, that is not located in N-well (the source and drain regions of an NMOS), or the active area where a gate does not cross over, that is located in N-well (the N-well contact region), with these areas enlarged by 0.2 μm."

Thus, there exist mask-generating layers that are generated by logical operations on layers that are hand drawn by the designer. Before layout design, it is necessary to confirm in the layout manual which layers are hand drawn by the designer and which layers automatically generate which other layers through which logical operations. In reality it is likely that a layout is started based on a reference inverter layout, but these things still need to be checked at least once.

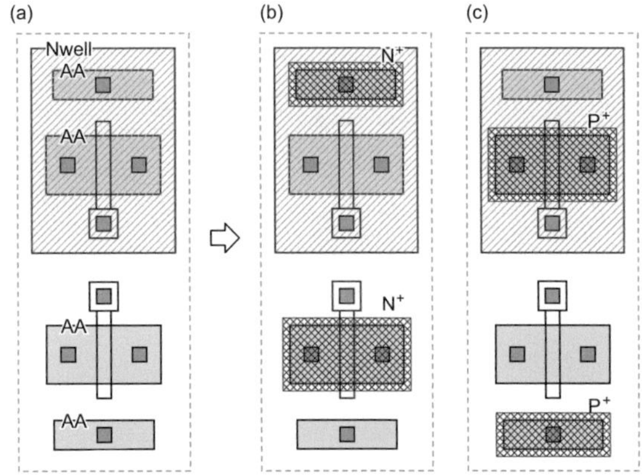

Fig. 3.17 Layer automatic generation, (**a**) layout, (**b**) automatic generation of N^+, (**c**) automatic generation of P^+

3.3 Basic Layout

3.3.1 Layout of Transistors

If we tell a layout novice to "draw an inverter layout," the circuit diagram of Fig. 3.18a may be directly realized as the layout in Fig. 3.18b (my very first layout was like this as well). Besides the lack of the body terminal (well tap), this may not exactly be a mistake. However, it is better to rotate the transistor to be vertical as in Fig. 3.18d. This may initially seem like a matter of personal layout style because that shown in Fig. 3.18b, d will function properly as transistors, but in a NAND gate layout, it is obvious that presented in Fig. 3.18g is better than that in Fig. 3.18f.

3.3.2 Layout of Resistors

"Regular" digital circuits seldom use resistors, but in analog circuits, resistors can become necessary.

In LSI, gate poly is used for resistors. Because these are not transistors, they are formed on STI regions rather than active regions. There are four types of gate poly as indicated in Fig. 3.19. The resistance value changes depending on whether the gate poly is doped with P^+ or N^+. Also, normally the transistor gate surface is silicided to reduce the gate resistance, but as in Fig. 3.19c, d, it is possible to bypass this procedure to use the gate poly as a high-resistance element. This is called "silicide protection," and a dedicated layer for this is used (this requires an additional mask which increases the process cost).

The resistance value of Fig. 3.19a, b is determined mostly by the silicide resistance and as a rule of thumb is approximately $10\,\Omega/\Box$. In Fig. 3.19c, d, the

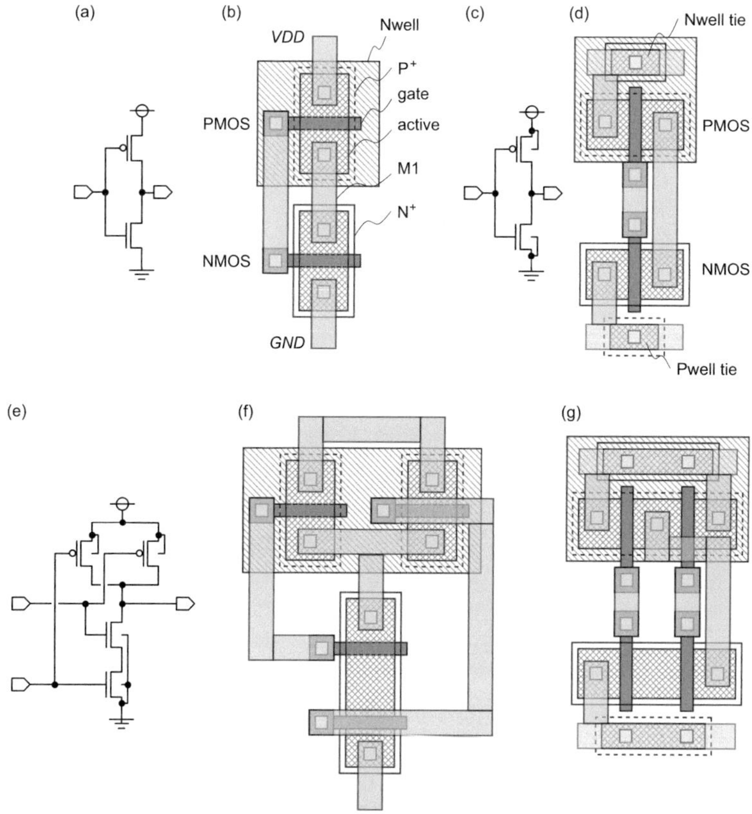

Fig. 3.18 CMOS layout. (**a**) Novice circuit diagram, (**b**) novice layout, (**c**) correct circuit diagram, (**d**) correct layout, (**e**) correct NAND circuit diagram, (**f**) novice NAND layout, (**g**) correct NAND layout

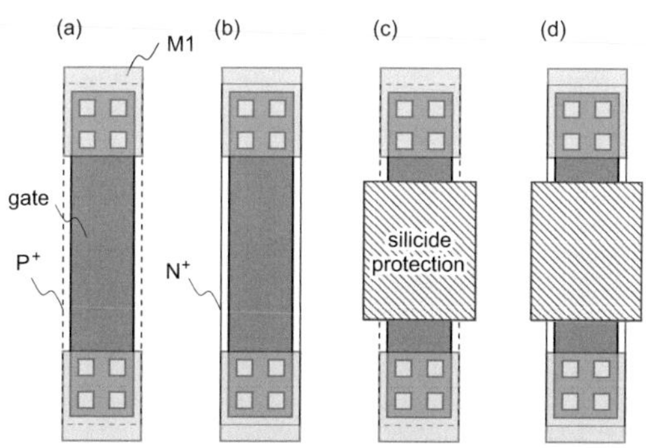

Fig. 3.19 Layout of resistors, (**a**) P^+ poly resistance, (**b**) N^+ poly resistance, (**c**) P^+ poly resistance no silicide, (**b**) N^+ poly resistance no silicide

unit resistance is several hundred Ω/□. The specific values for these should be prescribed in the layout manual.

It is worth mentioning that if the resistance value can be a function of voltage and the accuracy is not of great importance, for example, in some rough LPF, then using a transistor ON resistance is probably the easiest way to achieve a resistor.

3.3.3 Layout of Capacitors

Capacitances in normal circuits slow down the operation speed and increase the power consumption. However, capacitors are necessary in some analog circuits. Practically implementable capacitors inside LSI are as follows, and the appropriate one must be chosen according to the application:

MIM capacitors Metal-insulator-metal (MIM) capacitors are not formed within regular layers, but rather a layer called the MIM layer is inserted as in Fig. 3.20a to create a structure that looks like that shown in Fig. 3.20b, forming a capacitor whose capacitance is independent of voltage. Normally, the MIM layer is inserted just above the second top layer. The capacitance per unit area value should be prescribed in the design manual.

MOS gate capacitors The gate oxide thickness is less than the plate separation of a MIM capacitor, and the capacitance per unit area is the largest for these capacitors. However, the capacitance value changes with the bias voltage. For example, the capacitance value shrinks if the channel is not formed. This can be seen as a type of varactor. The NMOS gate capacitor of Fig. 3.20c and the PMOS gate capacitor of Fig. 3.20d are connected in

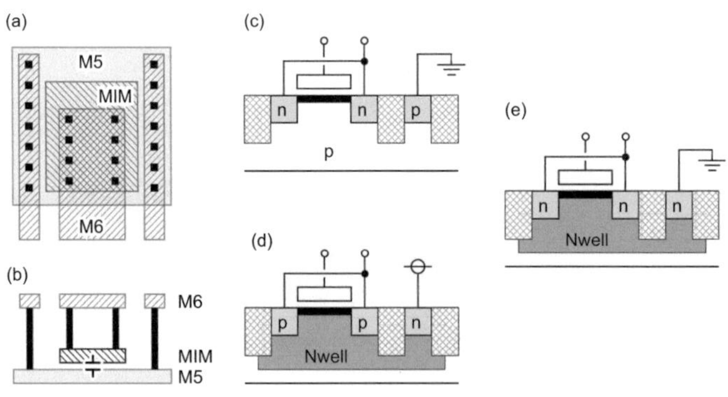

Fig. 3.20 Layout of capacitors, (**a**) MIM capacitors, (**b**) cross section of MIM capacitors, (**c**) NMOS gate capacitors, (**d**) PMOS gate capacitors, (**e**) varactors

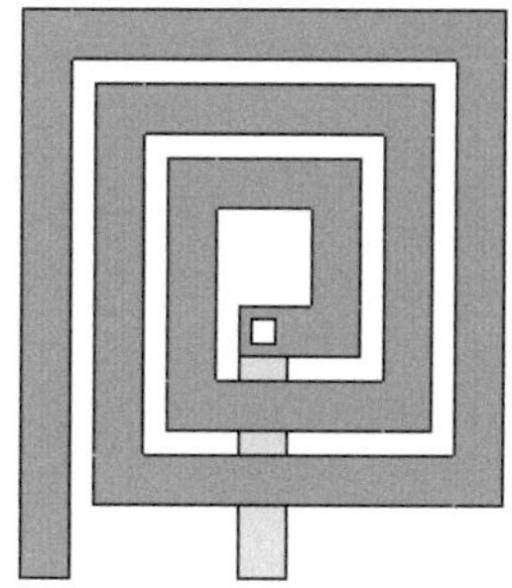

Fig. 3.21 Layout of inductors

parallel to reduce the dependence of the capacitance on the bias voltage, as necessary. Also, in processes below 90 nm with significant gate leakage currents, it is necessary to be wary of the fact that there will be some current flow between the two terminals of the capacitor. The basic operation is that of a MOS transistor, and thus the capacitance value can be determined through SPICE simulations.

Varactors

Varactors, or variable capacitors, change their capacitance with voltage. In the example in Fig. 3.20e, an NMOS is formed on N-well, and as the channel formation changes with the gate voltage, a capacitance that depends on the voltage is realized. These can be treated as special MOS transistors, and NMOS models such as BSIM3 are somewhat forcefully used in simulation

3.3.4 Layout of Inductors

An inductance is formed when wiring is laid out in a spiral, as in Fig. 3.21. A larger spiral becomes a larger inductance, but at the same time, the resistance as well as capacitance with the substrate will increase, degrading the inductor characteristics. Designers must put in a lot of effort to layout good inductors, such as to use the top most layer or to use multiple layers to reduce the internal resistance. As a rule of thumb, a $100 \times 100\,\mu$m spiral becomes roughly several nH of inductance.

3.4 Layout Editors

The layout editor is the CAD tool used to conduct layout design. Basically it can be thought of as a "simple drawing tool."

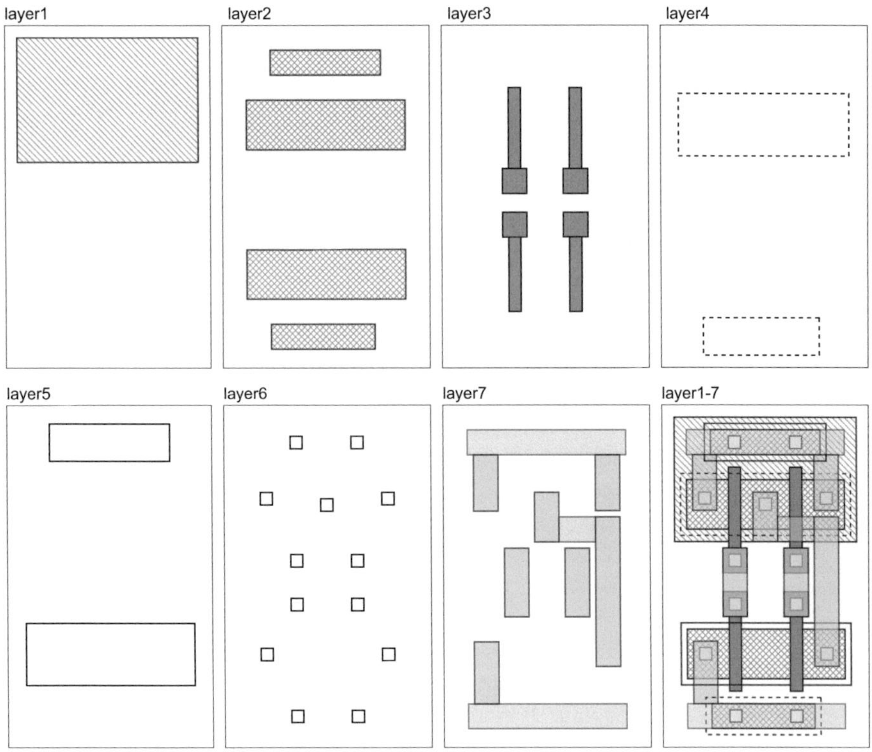

Fig. 3.22 Layers

3.4.1 *Layers*

The layout editor has the "layer" concept, which corresponds to the mask in an actual fabrication process. As shown in Fig. 3.22, the pattern for each mask is drawn in each layer. Within the layout editor itself, there is no need to limit layer 1 to N-well, layer 2 to active area, and so on. It is up to the designer in the end to specify this correspondence, such as that "layer 1 is for the N-well mask," in a setup file. The layout editor has many useful functionalities, such as specifying the layer for the pattern being drawn or turning on and off the visibility of each layer, as in showing layer 1 only or showing all layers 1–7.

3.4.2 *Display and Grid*

To make the three-dimensional structure of the transistors and interconnect easy to visualize, the display color and pattern for each layer can be specified. Just by changing these colors, it can become difficult to "see" the circuit in layout, and it is possible for seniors who are checking your design to tell you "the color settings are different from mine so I cannot understand your layout." Initially, you should use the color settings used traditionally in your group. A color setting file is used by layout editors, so if you want to use an original color setting, a file for this as well as a color setting file for showing your layout to others should be prepared.

Also, as shown in Fig. 3.13, the minimum grid is set for each layer, so the minimum grid in the layout editor should be determined and all of the drawn objects should fall on the grid.

3.4.3 Objects

In most cases, the objects used in layout are given as below and as indicated in Fig. 3.23:

(a) **Rectangles**	A rectangle that is determined by the two endpoints of its diagonal
(b) **Paths (no extension)**	A straight line connecting two points. The width is specified separately. A default path width can be set for each layer.
(c) **Paths (with extension)**	A straight line connecting two points. The amount of extension can also be specified. However, the extension should be set to either zero or half of the path width. For hand layout, it is recommended that zero extension be used.
(d) **Polygons**	Each vertex is specified. Sometimes, each polygon is limited to a maximum of 128 vertices. Whether 45° angles and arbitrary angles are allowed should be checked in the design manual.
(e) **Overlapping rectangles**	Even if shapes overlap, the final shape is interpreted as the OR of the multiple shapes.
(f) **Cells**	Multiple objects can be combined into a cell and treated as one block. For example, the shape in

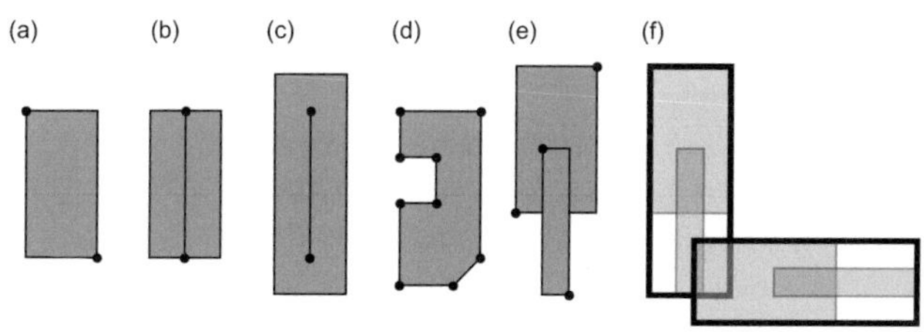

Fig. 3.23 Objects, (**a**) rectangle, (**b**) path, (**c**) path, (**d**) polygon, (**e**) overlapping rectangles

(e) can be turned into a cell and placed down as a cell. When the contents of a cell are edited, all instances of the cell are updated with the change.

Usually metal wires are drawn with paths, and all other shapes are drawn with overlapping rectangles.

3.5 Layout Know-How

3.5.1 Layout Editor Settings

First the layout editor should be customized so that it is easy for you to use as a tool.

3.5.1.1 Color Settings

As mentioned in Sect. 3.4.2, colors that match your likings can be used. Beginners should use the same color setting file as those around you.

3.5.1.2 Grid Settings

As mentioned in Sect. 3.4.2, the grid should be set correctly. Also, layout can become easier if the grid is set to be larger than the minimum allowed grid. For example, even if 0.01 μm grids are allowed in the design rule, if the design can be done with a 0.02 μm grid or a 0.04 μm grid, then conducting design with a 0.04 μm grid is easier. Here, the grid settings can be changed depending on the situation, as in doing most of the design with a 0.04 μm grid but using a 0.01 μm only when necessary.

3.5.1.3 Dividing Layers

Even if they are the same M1, it is strongly recommended to change the display of supply lines, ground lines, and signal lines within the layout editor. Here, these layers are combined into one when outputting the GDSII file.

For example, as shown in Fig. 3.24, METAL1VDD is red with a right diagonal hatch, METAL1GND is red with a left diagonal hatch, METAL1SIG is red with a cross hatch, METAL2VDD is blue with a right diagonal hatch, METAL2GND is blue with a left diagonal hatch, METAL2SIG is blue with a cross hatch, and so on, thereby distinguishing the routing layers with the color and distinguishing between supply, ground, and signal with the hatch pattern. This may be difficult to appreciate

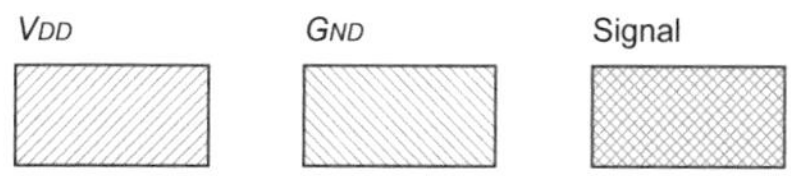

Fig. 3.24 Layer division for the same routing layer

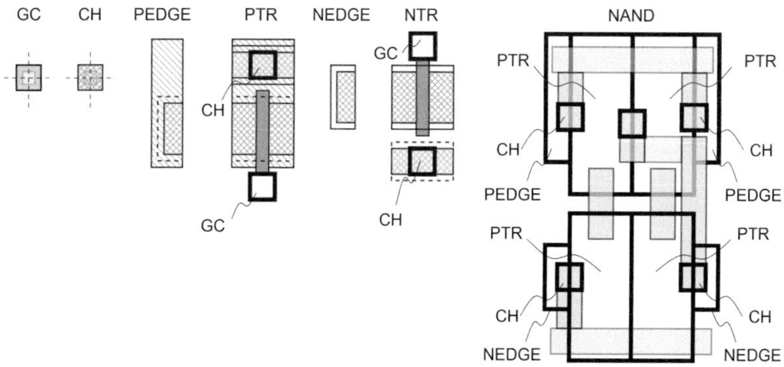

Fig. 3.25 Cells and hierarchical layout

for layouts of a small circuit such as a DFF, but this is definitely useful for routing supply lines at the chip level.

3.5.2 Hierarchical Layout

Layout is built by utilizing hierarchical design that brings together blocks which have been made into cells and are used repeatedly. For example, in the layout of Fig. 3.18g, drawing all of the squares one by one is cumbersome, and a hierarchical design should be conducted as indicated in Fig. 3.25. In this example, the cell GC is a cell with a gate, contact hole, and M1. The cell CH is a cell with a gate, active area, and M1. Transistors are also turned into cells PTR and NTR under the presumption that they will be connected in parallel, and the transistor edges are also turned into cells PEDGE and NEDGE. A NAND gate made in this fashion is also turned into a cell, and when creating a SRLATCH, NAND cells can be placed and connected with M1 and CH.

3.5.3 Double Back

Normally the PMOS is drawn on top and NMOS on bottom, as in Fig. 3.18. However, as the layout grows and transistors are placed in the vertical direction, as shown in Fig. 3.26, the layout becomes better if the layout is alternated with one

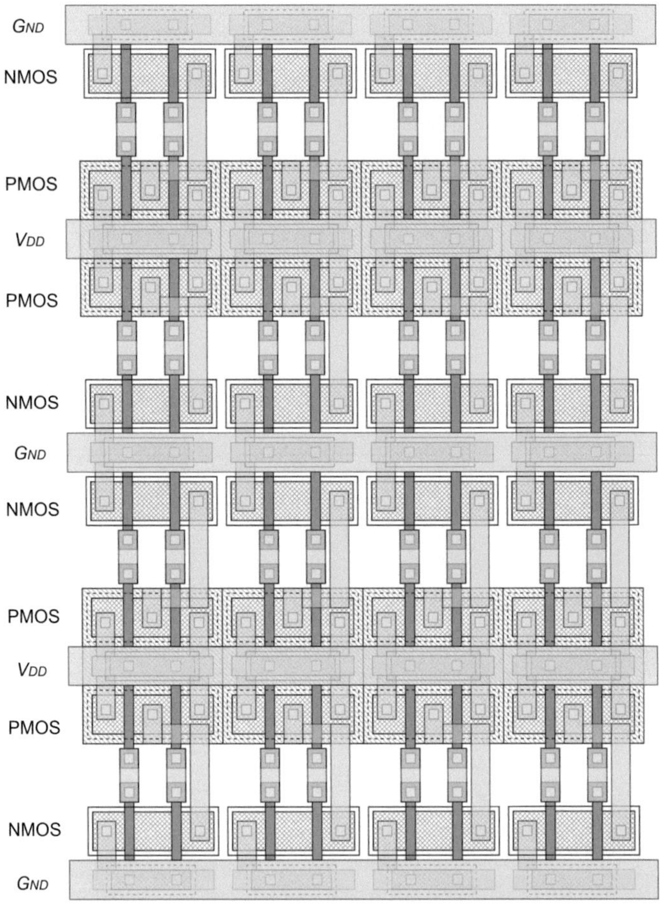

Fig. 3.26 Double back

where the NMOS is on top and the PMOS is on bottom, because supply lines can be shared and N-wells can be combined. This arrangement is called double back (in the figure the supply and ground lines are displayed as the same, but this will become easier to understand if as indicated in Fig. 3.24 the supply and grounds are displayed differently).

3.5.4 *Supply Lines*

Because a large current flows through the supply, the resistive component must be made as small as possible to avoid a voltage depression due to an IR drop. Several measures can be taken, such as making the routing width fatter, connecting several

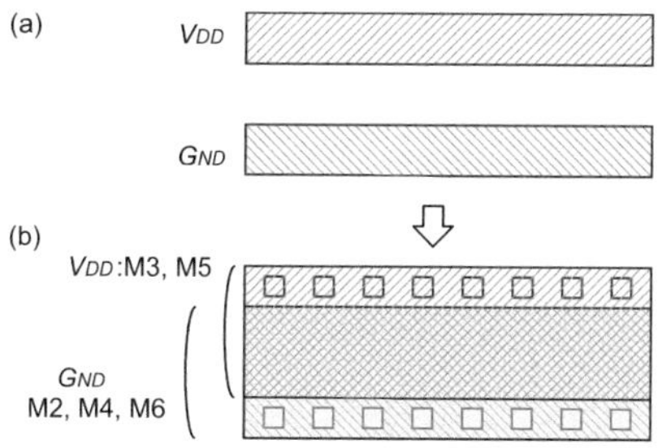

Fig. 3.27 Layout of supply lines

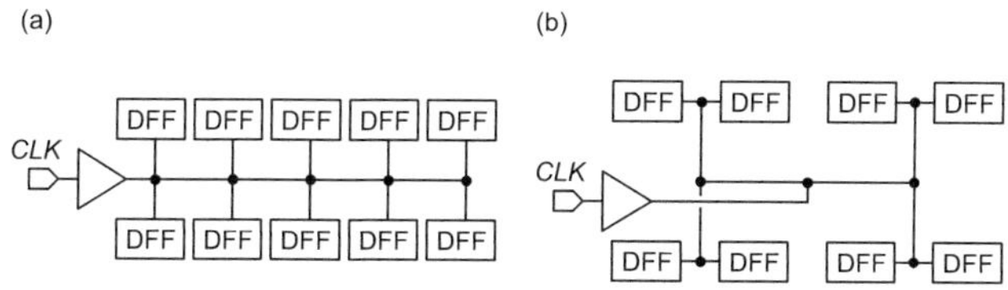

Fig. 3.28 Clock routing, (**a**) fish bone, (**b**) H-tree

layers of wiring together, using the upper layers more frequently for their thickness, and making the routing as short as possible. Also, the amount of voltage drop due to sudden currents can be reduced by placing capacitors between supply and ground. Measures such as trying to make supply and ground wires overlap as much as possible and placing capacitors in the spaces after final routing is finished are taken (Fig. 3.27). However, making supply and ground wires overlap too much will get in the way of trying to route signal wires so this should be done only to some moderate degree.

3.5.5 *Clock Distribution*

In synchronous circuits, it is desirable for all the clock signals for all DFFs to arrive at the same time. By considering the CLK wiring delay and the delay due to buffer insertion, the fish bone structure of Fig. 3.28a and the H-tree structure of Fig. 3.28b are combined to distribute the clock signal to the entire chip.

When the digital circuit contains hundreds of thousands of gates, obviously an exclusive CAD tool is used to lay down the clock tree. However, the clock is, along with the supply lines, the most important routing, and some portions, such as the clock synchronization between blocks, can become manual labor.

Also, in cases such as shift registers in Fig. 3.29a, where there is little logic between DFFs, it is possible for data to "pierce" the DFFs if the clock signals do not arrive at the same time. For example, in Fig. 3.29a, at time t_n the first and second

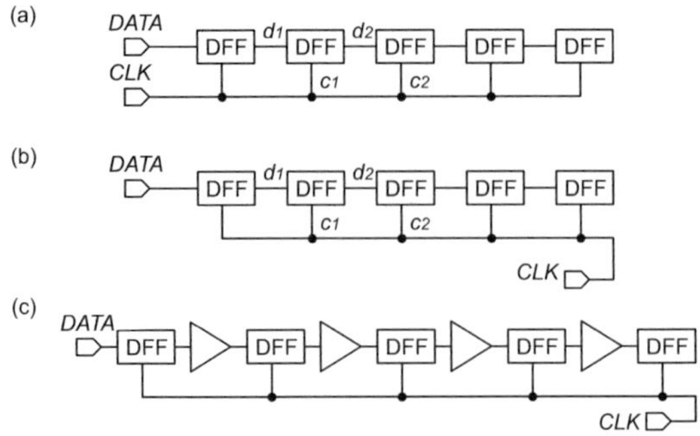

Fig. 3.29 Shift register wiring. (**a**) Regular circuit diagram, (**b**) circuit diagram with layout considerations, (**c**) circuit diagram with buffer insertion

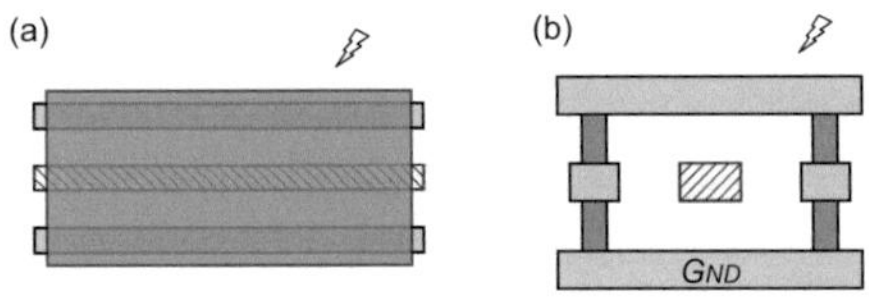

Fig. 3.30 Shield for important signal routing. (**a**) Layout, (**b**) cross section

DFFs are outputting data d_1 and d_2; if c_1 goes high before c_2, then the second DFF would output d_1 due to c_1 rising; and then when c_2 rises, the third DFF would also output d_1. To avoid this data feedthrough, the clock can be supplied from the right as shown in Fig. 3.29b to make sure the clock arrives at c_2 before arriving at c_1. On the circuit diagram, the netlist will be the same whether the clock is supplied from the left or the right and the simulation results will be the same, but a circuit diagram that has layout taken into consideration at the circuit design step will reduce the number of errors during layout. To make matters even safer, buffers can be inserted between DFFs as shown in Fig. 3.29c to buy some time.

3.5.6 *Shields*

In analog design, there are wires which can get in trouble if it receives the effects of noise, such as the wires that carry important analog values. In these cases, as shown in Fig. 3.30, the wire should be surrounded by G_{ND} routing. By doing so, even if noise comes flying, the G_{ND} routing shields the signal and the signal does not get affected by noise. However, this will add routing capacitance from signal to G_{ND}, so the trade-off with increased power dissipation or reduced operating speed must be taken into consideration.

3.6 Layout Verification

Layout data that has been designed with a layout tool is output in a GDSII (read G-D-S-two) industry standard format, which then is used in a variety of verification tools. The data file output in the GDSII format is called the "GDS file" or the "stream file."

The layout objects included in the GDS file are a collection of simple shapes like those in Fig. 3.23. Also, the shapes are on a 1nm scale in the GDS file, and all of the layout data are represented with integers with no decimal point.

3.6.1 DRC

In Design Rule Check (DRC) verification, whether the layout meets the rules prescribed in the design manual, such as those indicated in Figs. 3.11 and 3.12, is checked.

3.6.1.1 Rule Files

The rules prescribed in the design manual are described in varying formats for each verification tool. The rule file is divided into the header and main portions. The header portion lists thing such as:

- the stream file name and the cell names inside the stream file for which DRC must be run
- formats and file names for run log and error outputs
- the grid size

and the main portion describes the rules themselves. For example, a file may look like this:

```
CONT     =  06
M1       =  08
ENC[tos]  CONT    M1     lt   0.1 output ERe06 35
```

and this means:

```
The sixth layer (CONT) must be enclosed by an extension
of at least  0.1um of the eighth layer (M1). If there are
sections less than 0.1um,  output to the ERe06 cell as
the 35th layer.
```

The rule file description format depends on each company's DRC tool, but the contents are the same. Normally, there is no need for the designer to understand and follow the main portion of the rule files, but rather understand the design rule manual shown in Fig. 3.12 and abide by these rules in layout and fix the layout where errors are indicated by the DRC tool.

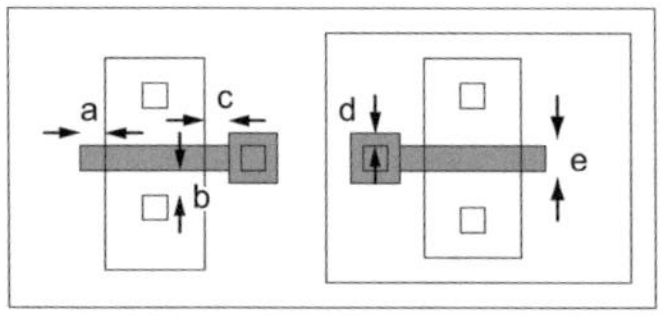

a	Min. extension of GATE beyond AA	0.2um
b	Min. clearance from CH to GATE	0.4um
c	Min. clearance from AA to GATE	0.1um
d	Min. extension of GATE beyond CH	0.1um
e	Min. width of GATE	0.2um

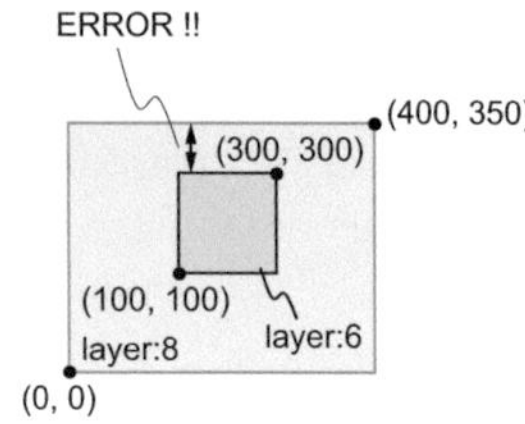

Fig. 3.31 DRC error example

3.6.1.2 Methods

Within the GDSII file, rectangles are represented by the upper left and lower right points, and paths are indicated by the start and endpoints and the path widths. Thus, the checks for whether the above mentioned rules are satisfied are nothing but integer operations on the figures. For example, as in Fig. 3.31, the operation would be "Between the eighth layer (0, 0), (400, 350) rectangle and the sixth layer (100, 100), (300, 300) rectangle, the sides and bottom are surrounded by 0.1 μm of extension so there is no problem. The top only has 0.05 μm of extension so this is an error." The calculations conducted here would be "make sure that (100, 100) is inside (0, 0) and (400, 350), and (300, 300) is also inside (0, 0) and (400, 350). For extensions in the x axis direction, the left is $100 - 0 = 100$ which is 0.1 μm, the right is $400 - 300 = 100$ which is 0.1 μm, and the bottom is $100 - 0 = 0.1$ μm so there is no problem. The top extension is $350 - 300 = 50$ which is only 0.05 μm so this is an error." Thus, the calculations are simple integer additions and subtractions.

3.6.2 LVS

In Layout Versus Schematic (LVS) verification, the layout data and the circuit diagram (netlist) are verified to be equivalent.

Similar to DRC, the circuit diagram and netlist are compared by recognizing the transistors and wiring connections from the collection of rectangles described in the GDS file. The program implements the same thing as a person following a layout and comparing with the circuit.

For example, a rule file might look something like this:

```
NWELL  =  31
ACTIVE =  01
PIMP   =  14
NIMP   =  15
POLY   =  05 TEXT = 05      ATTACH=POLY
CONT   =  06
```

Fig. 3.32 Inverter LVS

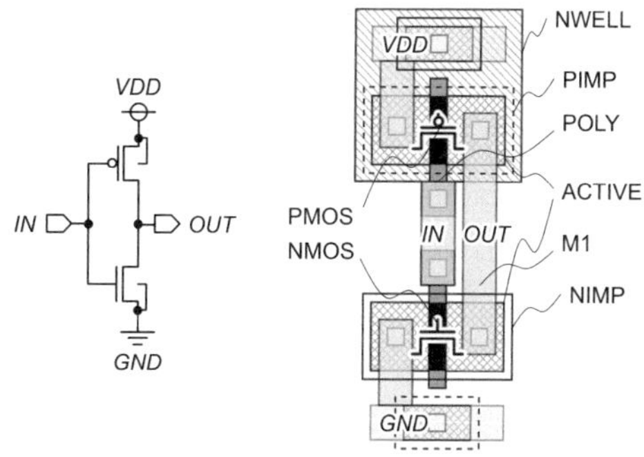


```
M1      =  08

and ACTIVE PIMP PREGION
and PREGION POLY PGATE
not PREGION PGATE PSD
element    MOS[P]   PGATE POLY PSD NWELL

and ACTIVE NIMP NREGION
and NREGION POLY NGATE
not NREGION NGATE NSD
not bulk NWELL PSUB
element    MOS[N]   NGATE POLY NSD PSUB

connect    M1 POLY by CONT
connect    M1 PSD by CONT
connect    M1 NSD by CONT
```

After specifying the relationships between the file layer number and the actual mask layer, the overlap between ACTIVE and PIMP is defined as PREGION, the overlap between PREGION and POLY is defined as PGATE, the portion of PREGION that is not PGATE is defined as PSD, and the PGATE region is defined as PMOS with POLY as the gate terminal, adjacent PSD as the source and drain terminals, and the overlapping NWELL to be the body terminal. Also, M1 and POLY are connected by CONT, M1 and PSD are also connected by CONT, and M1 and NSD are also connected by CONT.

These kinds of rules that define the method of transistor recognition as well as the connectivity relationship from layer to layer allow the LVS tool to recognize transistors and their connections, compare with a netlist, and determine whether the layout is correct or not and if not specify what is incorrect. That is, the layout and circuit diagram in Fig. 3.32 are checked to see if they match. Not surprisingly, there is a need to specify the locations of terminal names (in this example *IN*, *OUT*, *VDD*, and *GND*). The terminal names are input with text called a "label," and in actual mask fabrication, all of the text is ignored.

There are many options when running LVS, and these options can be enabled or disabled depending on the design, so a consensus should be reached within your group before starting layout. As shown in Fig. 3.33, some typical options to be wary of are given below. (The option names below are called different things in

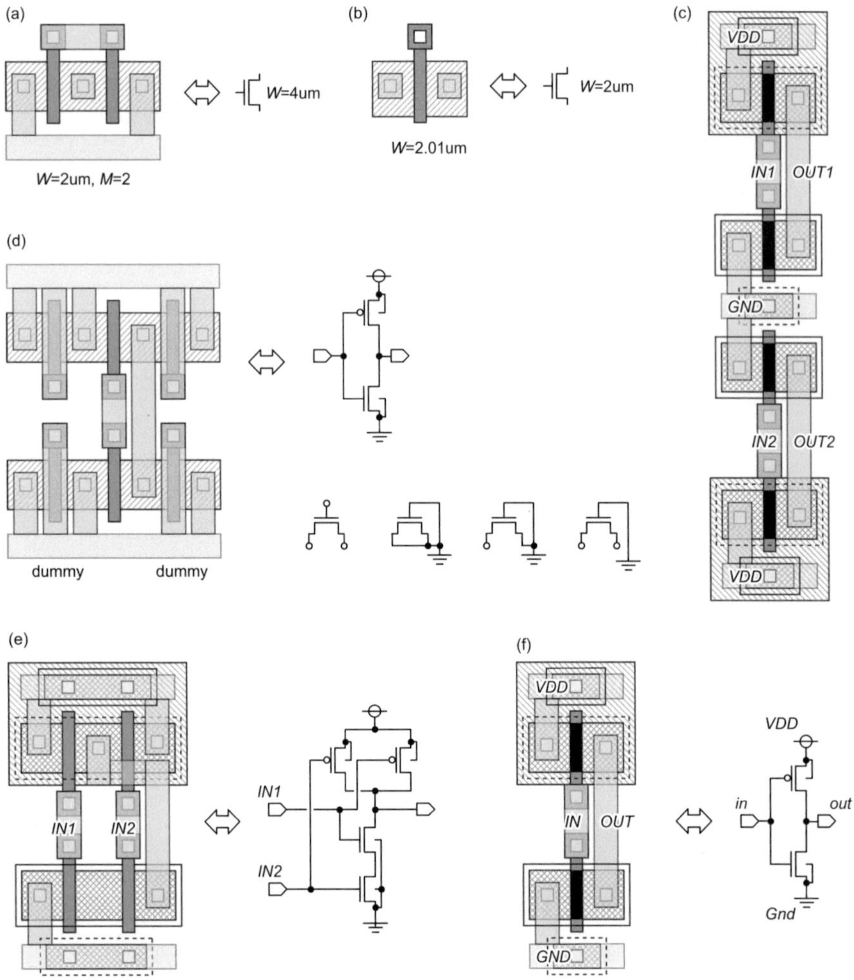

Fig. 3.33 LVS options

each company's tool, so find the options that sound similar within the LVS rule and confirm.)

(a) **Merge parallel** For example, whether to consider two $W = 2\,\mu$m transistors in parallel and a single $W = 4\,\mu$m transistor as matching or not.

(b) **Property** The tolerable error range for L and W. The transistor size is extracted from the layout, but this option can be set so that LVS will consider it a match if, for example, the layout size differs from the size given in the netlist by less than 5 %.

(c) **Virtual connect** For example, with *VDD*, whether to allow nodes to be considered connected by giving them the same label even though they may not be connected in layout.

(d) **Filter** For example, in the case of an NMOS, this is the option to ignore the device if the gate, source, and drain are all open; if the gate, source, and drain are all connected to *GND*; if the gate and either the source or drain are connected to *GND*; if the gate is connected to *GND*; and so on. As a common example, this option determines whether dummy transistors placed in layout for variation mitigation must also be drawn in the circuit diagram or not.

(e) **Recognize gates** For example, if the inputs to a two-input NAND gate are *IN*1 and *IN*2, whether to pass the check even if *IN*1 and *IN*2 are interchanged (in the netlist the NMOS on the *OUT* side is *IN*1, but in layout the *GND* side is *IN*1), because this is a two-input NAND gate, so the logic is the same.

(f) **Case sensitivity** For example, whether to consider *IN*1 and *in*1 to be the same. It is rarely the case that we name separate terminals *IN*1 and *in*1 within the same circuit, but rather this option is to check whether to match a *IN*1 in the netlist and a *in*1 in layout (or vice versa).

3.6.3 ERC

In Electrical Rule Check (ERC) verification, whether the circuit is properly electrically connected or not is checked. For example, some checks are listed below:

- N-well is connected to the supply and P-well to ground.
- There is a path through source/drain to the supply or the ground.
- The distance from the gate of each transistor to the body contact is below a certain threshold.
- The terminals of transistors are not floating and connected to somewhere.

However, ERC generates many pseudo-errors, and a careful judgment is necessary on whether the error is real or ignorable. For example, in Fig. 3.34a, the *GND* side PMOS body is tied to the output and not *VDD* on purpose, so this is not an error. In Fig. 3.34b the *IN* terminal is only connected to gates and there are no paths to supply or ground, but this node will be connected to a transistor drain in the previous stage, so this is also not an error. In Fig. 3.34c if the distance between the transistor body terminal and body contact is long and the resistance increases a latch-up will occur, so this is a true error (in reality, a distance of tens of μm is prescribed to be tolerable, and a separation on the order indicated in the figure is not a problem).

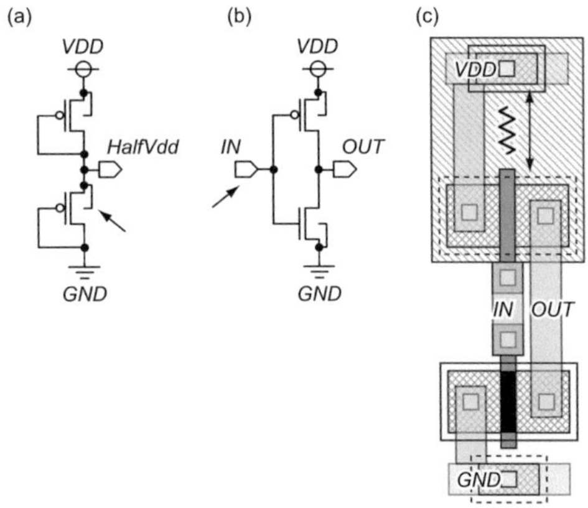

Fig. 3.34 ERC

If each transistor is drawn in the circuit diagram and then drawn in layout, as long as you are careful of the body contact distance, a passing LVS in most cases automatically satisfies ERC.

3.6.4 Antenna Check

Compliance to the antenna rules explained in Sect. 3.2.5 is checked. This occurs most often in the connection between IO and core circuitry.

3.6.5 Density Check

Compliance to the density rules explained in Sect. 3.2.3 is checked. This is rarely a problem because dummies are automatically generated on the fabrication side after the designer submits the final layout. However, if dummies must be generated on your own, then you must also conduct the density check.

3.6.6 Verification Types and Order

In the end LVS, DRC, ERC, antenna, and density must all be satisfied. Recently there are sometimes more detailed checks, but they are usually a subset of one of the verifications mentioned above. For example, with the grid as explained in Sect. 3.2.2, if there is a need to separately check whether the shapes are on the defined grid, then this check can be included in DRC.

LVS and DRC are the most basic, and during design, these two checks only are conducted. After some sizable portion has been finished, the ERC and antenna checks, and density if necessary, are conducted. Whether to run LVS or DRC first is a personal preference or taught from seniors and can vary, but I personally like to run LVS first.

3.6.7 Flat Verification and Hierarchical Verification

For example, if the same inverter circuit is cascaded as in Fig. 3.35, running LVS and DRC on all transistors and all layout data as in Fig. 3.35a is called flat verification. On the other hand, if LVS and DRC are run on the inverter cell first, then LVS and DRC are run on the cell interconnect and boundaries, and so on as in Fig. 3.35b; this division into a hierarchy is called hierarchical verification. In this example it makes little difference whether the verification is flat or hierarchical, but to verify an entire SRAM chip with a hundred million identical memory cells, using hierarchical verification will remarkably improve the execution speed. The same holds true for a logic circuit which contains DFFs in large quantities. In this kind of verification, the execution time and memory usage are proportional to the square of the number of elements N. Thus, even in a circuit with few repetitions, circuits that can be divided

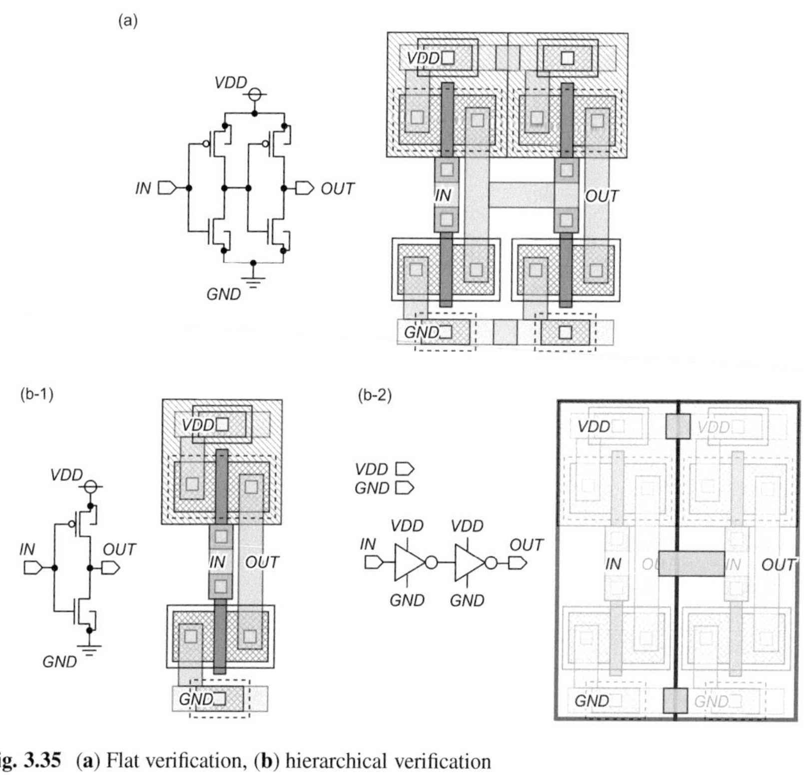

Fig. 3.35 (**a**) Flat verification, (**b**) hierarchical verification

easily should be divided for verification, and the connections should be verified separately for a faster run time. Also, hierarchical verification is well suited for multicore execution, because it divides the verification portions.

Most verification tools allow the use of hierarchical verification options, and its active use is recommended.

Chapter 4
Interconnect RC Extraction

Finally, you've finished layout. LVS and DRC also pass. After a sigh of relief, you're told to "run post-extraction simulations," and it seems that the circuit designed for 5 GHz only runs at 3 GHz. What should you do...?

4.1 Parasitic Resistance and Parasitic Capacitance

The cross-sectional image of a transistor and interconnect is shown in Fig. 4.1a. There is resistance in the interconnect and via that connect transistors and capacitance within the interconnect and between the interconnect and G_{ND}. In the two-stage inverter circuit layout of Fig. 4.1b, adding the interconnect resistance and capacitance will turn the circuit into what is shown in Fig. 4.1c. Because these are not purposefully placed gate poly resistors or MIM capacitors by the designer but rather resistances and capacitances that come about against the designer's will, they are called parasitic resistance and capacitance, interconnect resistance and capacitance, or interconnect RC. Also, the extraction of these parasitic resistances and capacitances is called RC extraction, interconnect RC extraction, or layout parasitic extraction (LPE).

As a result of the interconnect RC extraction, the circuit characteristics will change (usually degrade) relative to the case where the interconnect RC is not taken into account, and the maximum operational frequency will decrease or the amplitude will decrease. Of course, this is closer to the actual LSI operating condition, and the degradation of circuit characteristics should not be lamented. In fact, this should be welcomed in the sense that the actual operation can be predicted at the design stage, and the circuit can be adjusted accordingly.

T. Nakura, *Essential Knowledge for Transistor-Level LSI Circuit Design*,
DOI 10.1007/978-981-10-0424-7_4

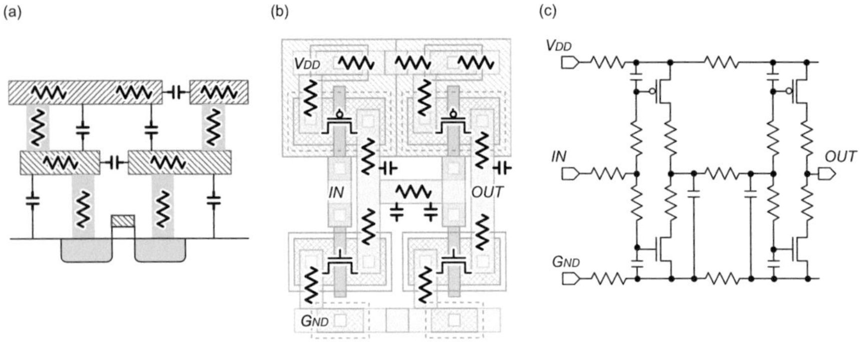

Fig. 4.1 (**a**) Cross-section of a transistor and interconnect, (**b**) layout of two inverters, (**c**) the actual circuit

4.2 Principles of RC Extraction Tools

4.2.1 Resistance Extraction

If a material has resistivity ρ $[\Omega \cdot m]$ as indicated in Fig. 4.2a, the resistance from end to end is given by

$$R = \rho \frac{l}{wd} = \frac{\rho}{d} \cdot \frac{l}{w}. \tag{4.1}$$

However, ρ and d are values determined by the fabrication process and the designer cannot change these. The designer can design l and w, and the resistance is proportional to l/w. The resistance when $l = w$, $\rho/d = R_{\square}$ $[\Omega/\square]$ is called the sheet resistance and represents the resistance per unit square (when viewed from the top). For example, in the wiring of Fig. 4.2b, there are five squares, so the total resistance is $5R_{\square}$. Also, a perpendicular bend as in Fig. 4.2c is said to have a resistance of $0.56R_{\square}$, and so this wiring has a total resistance of $5.56R_{\square}$.

For contacts, there are the contact hole resistance and gate contact resistance for the connection between transistor terminals and M1, as well as the via resistance for the connections in metal interconnect.

In an interconnect RC extraction tool, each routing layer is given a sheet resistance value, and the actual interconnect resistance is extracted from the layout data. For example, if the wiring width is doubled, the wiring resistance is halved. For contact resistances, a resistance value is given for each contact or via, and the actual resistance is extracted from the layout data. For example, if two contacts are placed, then the contact resistance is halved.

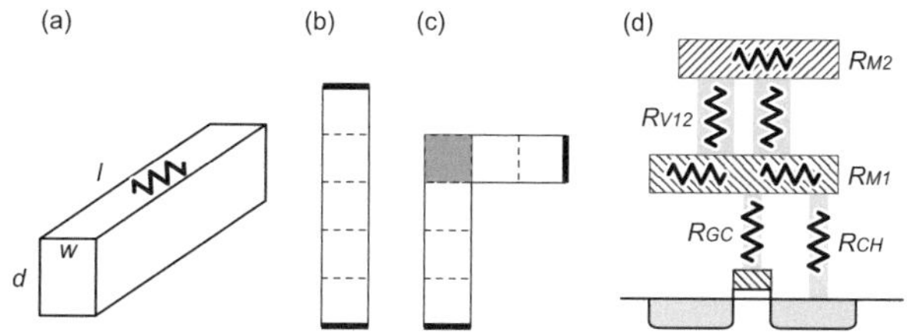

Fig. 4.2 (**a**) Wiring, (**b**) $5R_{\square}$, (**c**) $5.56R_{\square}$, (**d**) cross section

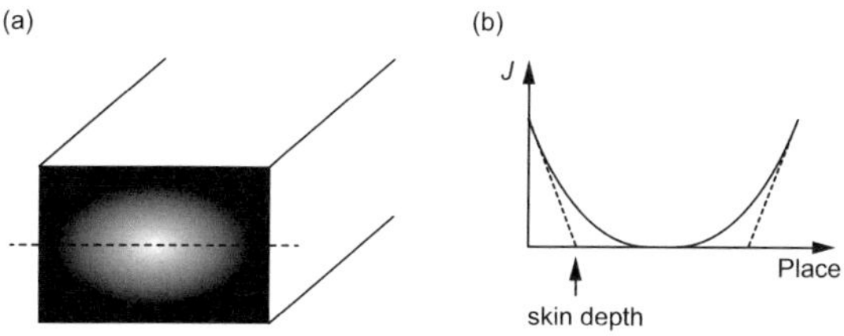

Fig. 4.3 Skin effect

4.2.1.1 Skin Effect

When a high frequency flows through a conductor with some resistance, the current tends to flow at the surface of the conductor as indicated in Fig. 4.3a. This is called the skin effect. In general, the current density J can be expressed as

$$J \propto e^{-\sqrt{\frac{\omega\mu}{2\rho}}x} \tag{4.2}$$

and the density decays exponentially with the depth. The depth corresponding to a e^{-1} decay of the current density

$$d = \sqrt{\frac{2\rho}{\omega\mu}} \tag{4.3}$$

is called the skin depth (Fig. 4.3b). The skin depth of copper at 1 GHz is about $2\,\mu$m which means the current can be thought to be evenly distributed within the wire, and thus the skin effect is usually not considered. However, when thinking about operation at tens of GHz, or for board wiring, the skin effect does need to be considered. Also, if the skin effect is considered and the resistance value becomes dependent on the frequency, then this means that SPICE simulations will take more time.

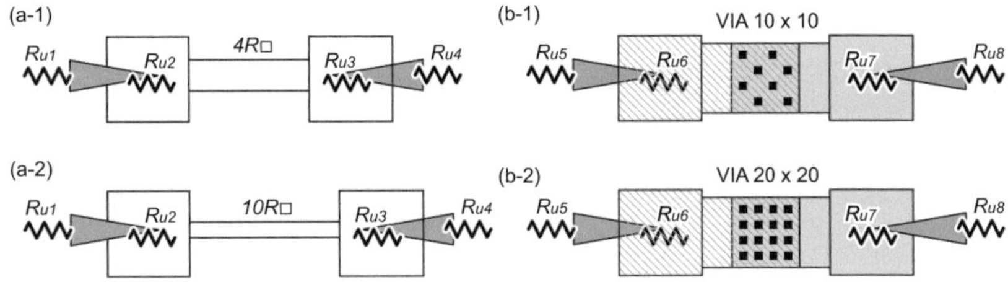

Fig. 4.4 (**a**) Sheet resistance measurement, (**b**) via resistance measurement

4.2.2 Resistance Measurement

In interconnect RC extraction, the sheet resistance of each metal layer and the resistances of vias between metal layers must be given as the technology file. However, it is sometimes desirable to measure the resistance of something that has been actually fabricated, because the routing and vias can have complex structures and the total resistance can be difficult to calculate from the resistivity of the material. Here, although pads are probed for the resistance measurement, a measurement circuit as in Fig. 4.4 is fabricated to remove the effects of various resistances R_{u*}, such as the resistance from the measurement device to the probe and the resistance between the probe and pad. The total resistances in Figures (a-1) and (a-2) are $R_{a1} = R_{u1} + R_{u2} + 4R_{\square} + R_{u3} + R_{u4}$ and $R_{a2} = R_{u1} + R_{u2} + 10R_{\square} + R_{u3} + R_{u4}$, respectively, which implies $R_{a2} - R_{a1} = 10R_{\square} - 4R_{\square}$, and thus the sheet resistance can be determined as

$$R_{\square} = \frac{R_{a2} - R_{a1}}{10 - 4}. \tag{4.4}$$

Similarly for the via resistance measurements, $R_{b1} - R_{b2} = R_{VIA}/100 - R_{VIA}/400$ where the resistance per via is R_{VIA}. Thus, the resistance per via can be determined by

$$R_{VIA} = \frac{(R_{b1} - R_{b2}) \times 400}{4 - 1}. \tag{4.5}$$

4.2.3 Capacitance Extraction

To integrate more transistors per unit area, it is necessary to make wires thinner. To reduce the wiring resistance, the metal layers have been made thicker. It is good to know that as a result of this, as shown in Fig. 4.5a, the intra-layer capacitance is larger than the interlayer (M1 and M2) capacitance.

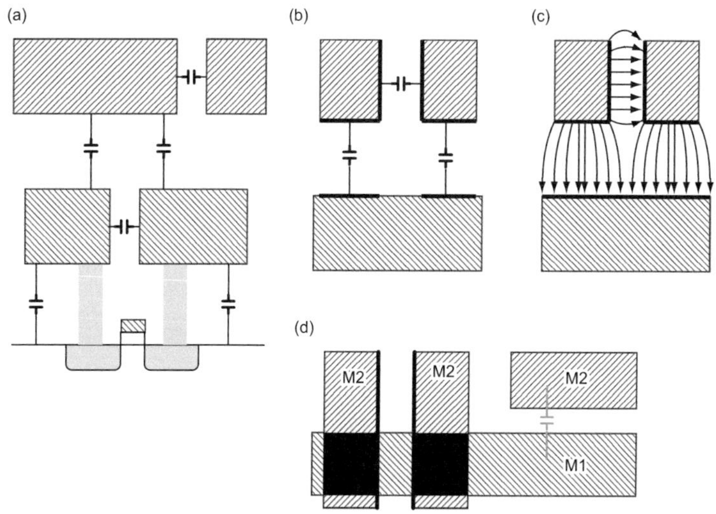

Fig. 4.5 Capacitance extraction

4.2.3.1 Parallel Plate Model and Edge Effects

In calculating the capacitance values, there is the method of using a simple parallel plate model ($C = \epsilon S/d$) as shown in Fig. 4.5b, and there is the method of extracting capacitance by considering edge effects as shown in Fig. 4.5c. In the parallel plate model, the fringe electric fields are ignored, leading to an extracted capacitance value much smaller than the actual value, and the model which considers the edge effects is more accurate. In addition, as in the view of the layout from the top shown in Fig. 4.5d, the parallel plate model recognizes the existence of capacitance in the overlap regions between M1-M2 and the adjacent portions between M2-M2, but when M1 and M2 are offset, the capacitance is not recognized.

4.2.3.2 Numerical Analysis

Because accurate capacitance extraction cannot be done with the parallel plate model, numerical analysis is used to calculate an accurate capacitance value which includes the edge effects. The basic formula uses Poisson's equation:

$$\frac{\partial^2 V}{\partial x^2} + \frac{\partial^2 V}{\partial y^2} + \frac{\partial^2 V}{\partial z^2} = -\frac{\rho}{\epsilon} \tag{4.6}$$

which is solved to determine V (here ρ is not the resistivity but the charge density); then the electric field is determined from

$$\boldsymbol{E} = -\nabla V \tag{4.7}$$

and electric charge Q is determined from

$$\int \boldsymbol{E} \cdot \boldsymbol{n} dS = \frac{Q}{\epsilon} \tag{4.8}$$

and finally capacitance C is determined from

$$Q = CV. \tag{4.9}$$

When the formulae become complicated, these cannot be solved analytically, so regions are divided into a fine mesh to calculate results numerically.

4.2.3.3 Difference Method

We will consider the two-dimensional case for simplicity. Just as with the oxide film, $\rho = 0$ when there is no charge in the region. When V in Eq. (4.6) is expressed as ϕ (In the field of numerical analysis with division into meshes, ϕ is used as the symbol for potential. The completed theory there is applied for capacitance extraction.),

$$\frac{\partial^2 \phi}{\partial x^2} + \frac{\partial^2 \phi}{\partial y^2} = 0. \tag{4.10}$$

When the region is divided as in Fig. 4.6, at the point ϕ_0,

$$\left(\frac{\partial^2 \phi}{\partial x^2}\right)_{\phi=\phi_0} = \frac{(\phi_3 - \phi_0)/h - (\phi_0 - \phi_1)/h}{h} = \frac{\phi_1 + \phi_3 - 2\phi_0}{h^2}. \tag{4.11}$$

When a similar expression is applied to the partial derivative in the y direction, Eq. (4.10) is

$$4\phi_0 = \phi_1 + \phi_2 + \phi_3 + \phi_4. \tag{4.12}$$

Namely, the relationship that the potential at a certain point is equivalent to the average value of the surrounding potential can be achieved. In the difference

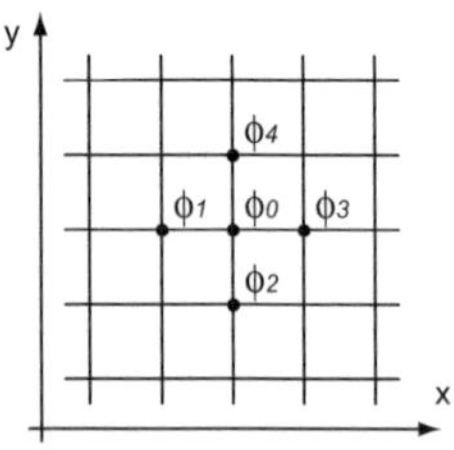

Fig. 4.6 Difference method

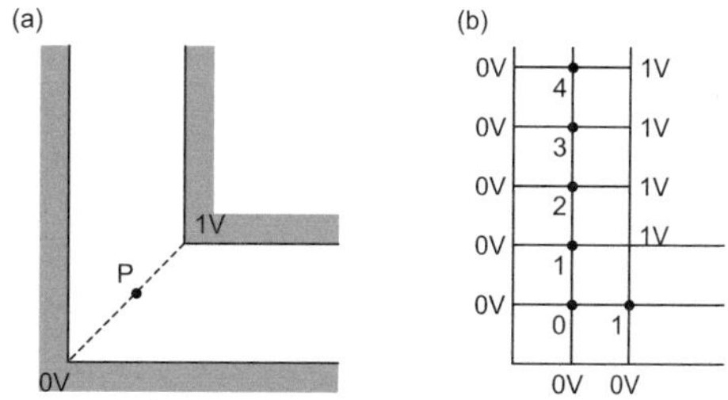

Fig. 4.7 Calculation of the potential distribution with the difference method

method, simultaneous equations like these that consider the boundary conditions are made for each divided point and then solved.

For example, to determine the potential distribution between two L-shaped conductors as in Fig. 4.7a, the region is divided into a mesh as in Fig. 4.7b, and the potential at point 4 ϕ_4 is assumed to be 0.5 V. When Eq. (4.12) is generated for points 3, 2, 1, and 0,

$$4\phi_0 = 2\phi_1,\ 4\phi_1 = \phi_0 + \phi_2 + 1,\ 4\phi_2 = \phi_1 + \phi_3 + 1,\ 4\phi_3 = \phi_2 + 1.5. \tag{4.13}$$

When these simultaneous equations are solved, we find that the potential at point P is $\phi_0 = 0.211$ V. The "correct" value is 0.202 V, which means that even with such a coarse mesh, a relatively close valuc can be attained. We can get to a more accurate value by dividing into a finer mesh.

4.2.3.4 Finite Element Method

Division into a mesh is also conducted in the finite element method, and each divided unit is called a finite element. The shape of a finite element can be anything, but it is common to use triangles. In the finite element method, the condition which minimizes the stored energy in each element is determined.

If an electric field E exists in a dielectric, an energy $\epsilon E^2/2$ per unit volume is stored, and in the finite element method, the electric field distribution which minimizes the sum of energies

$$W = \frac{1}{2}\int \epsilon E^2 dv \tag{4.14}$$

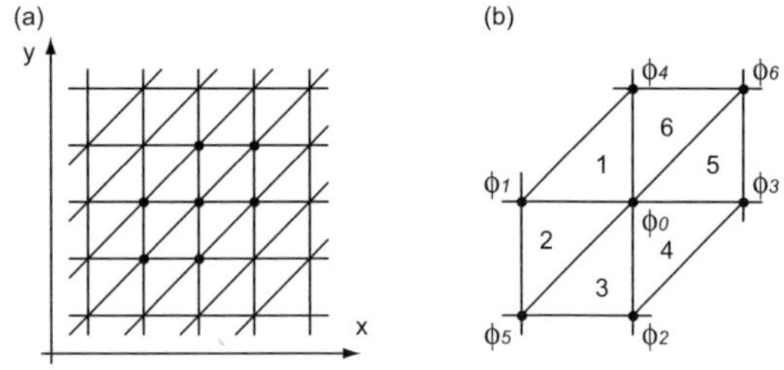

Fig. 4.8 Finite element method

is determined. We will consider the case when Fig. 4.6 is further divided diagonally and turned into a mesh as in Fig. 4.8. The energy stored in region 1 (W_1) is

$$W_1 = \frac{1}{2}\epsilon(E_x^2 + E_y^2)dv = \frac{1}{2}\epsilon\left\{\left(\frac{\phi_0 - \phi_1}{h}\right)^2 + \left(\frac{\phi_4 - \phi_0}{h}\right)^2\right\} \times \frac{h^2}{2} \tag{4.15}$$

$$= \frac{1}{4}\epsilon\{(\phi_0 - \phi_1)^2 + (\phi_4 - \phi_0)^2\}. \tag{4.16}$$

When the total energy of the six triangles is found by taking the sum

$$W_{1\sim6} = \sum_{i=1}^{6} W_i \tag{4.17}$$

$$= \frac{1}{2}\epsilon\{(\phi_0 - \phi_1)^2 + (\phi_0 - \phi_2)^2 + (\phi_0 - \phi_3)^2 + (\phi_0 - \phi_4)^2\}$$
$$+ \frac{1}{4}\epsilon\{(\phi_1 - \phi_5)^2 + (\phi_2 - \phi_5)^2 + (\phi_3 - \phi_6)^2 + (\phi_4 - \phi_6)^2\}. \tag{4.18}$$

Therefore, the ϕ_0 that minimizes the total energy must satisfy the following condition:

$$\frac{\partial W_{1\sim6}}{\partial \phi_0} = \epsilon\{4\phi_0 - (\phi_1 + \phi_2 + \phi_3 + \phi_4)\} = 0. \tag{4.19}$$

This equation is the same as Eq. (4.12) from the difference method.

We ended up solving the same equation for both the finite element method and the difference method, but the finite element method is based on the principle of minimum energy and thus more general, and the difference method is often considered one kind of the finite element method.

The analytical engine that conducts numerical calculations with division into meshes, as opposed to analytical equations such as the parallel plate model, is called a field solver.

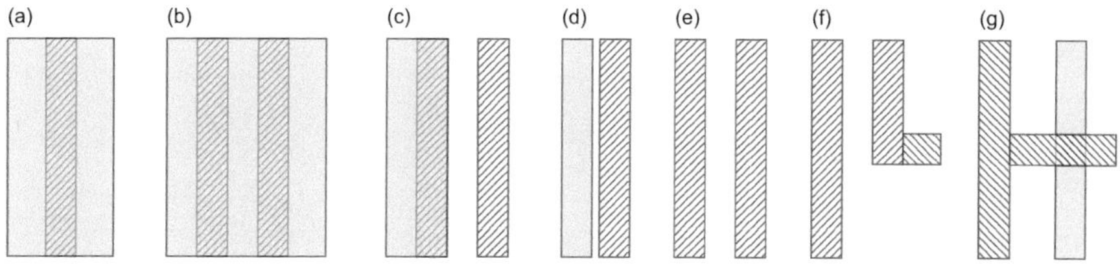

Fig. 4.9 Lookup table

4.2.3.5 Lookup Tables

An accurate capacitance extraction is possible by using a field solver, but if the scale of the circuit is large and a field solver is used on millions of wires, the calculations will never be finished. On the other hand, the error is too large with the parallel plate model. In an actual RC extraction tool, the lookup table method is often used. As shown in Fig. 4.9, capacitance values are calculated beforehand with field solvers for several typical patterns, shapes in the actual layout are matched against shapes in the tables, and based on the calculated capacitances, the capacitance of layout wiring is extracted. The basic shapes and the number of shapes stored in lookup tables vary from tool to tool.

4.3 AD/AS/PD/PS and HDIF

After interconnect RC extraction, the circuit in Fig. 4.10a turns into that in Fig. 4.10b with the addition of the RC of the wiring. Also, if a netlist without AD/AS/PD/PS is used, as in

```
m0 net1 IN   VDD VDD P L=180e-9 W=5e-6
m1 net1 IN   GND GND N L=180e-9 W=2e-6
m2 out net1 VDD VDD P L=180e-9 W=5e-6
m3 out net1 GND GND N L=180e-9 W=2e-6,
```

then this becomes

```
Cg1 M0:GATE 0 8.524e-17
C...

R1 M0:GATE IN 10.45
R2 M0:DRN NET1:1 10.28
R...

M0 M0:DRN M1:GATE VDD VDD P l=0.18u w=5u
+ ad=3.35p as=3.35p pd=11.34u ps=11.34u
M1 M1:DRN M0:GATE GND GND N l=0.18u w=2u
+ ad=1.34p as=1.34p pd=5.34u ps=5.34u
M2 M2:DRN M2:GATE VDD VDD P l=0.18u w=5u
```

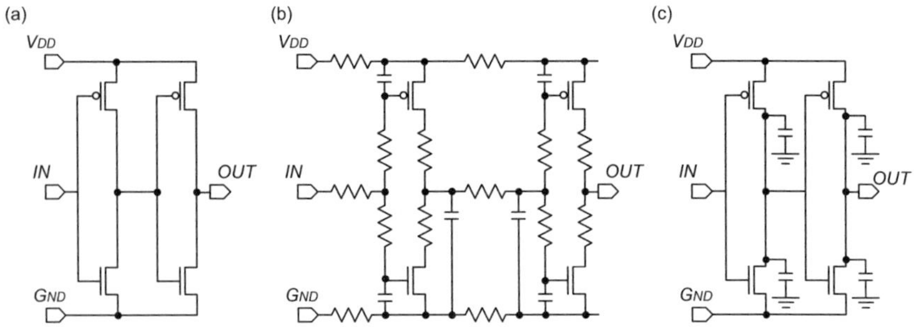

Fig. 4.10 Interconnect RC circuit diagrams for (**a**) before extraction, (**b**) after extraction, (**c**) overload

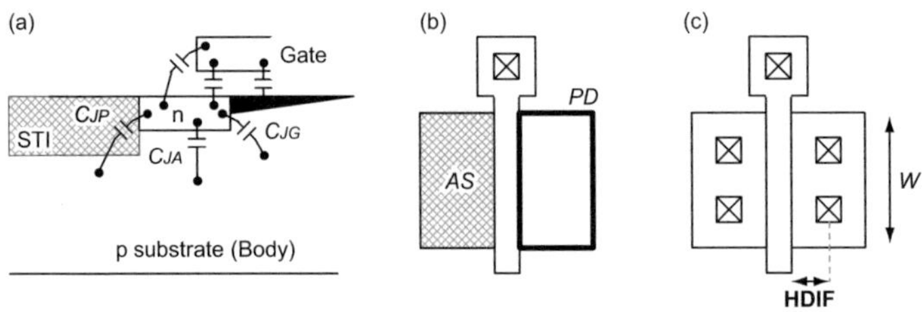

Fig. 4.11 Transistor capacitances

```
+ ad=3.35p as=3.35p pd=11.34u ps=11.34u
M3 M3:DRN M3:GATE GND GND N l=0.18u w=2u
+ ad=1.34p as=1.34p pd=5.34u ps=5.34u,
```

and AD/AS/PD/PS are added along with the interconnect RC. Here, the transistor capacitances are the gate and junction capacitances, as indicated in Fig. 4.11a, and AD/AS/PD/PS are values related to the junction capacitances. The drain junction capacitance is calculated as

$$C_D = C_{JA} \times AD + C_{JP} \times (PD - W) + C_{JG} \times W. \tag{4.20}$$

If the netlist has omitted AD/AS/PD/PS, these can be automatically calculated by assuming dimensions as in Fig. 4.11c, with models such as BSIM3 where the **HDIF** parameter is valid. However, if **HDIF** is not defined within SPICE parameters, or if the model does not consider **HDIF** such as with BSIM4, these parameters will be handled as AD = AS = PD = PS = 0 (but the gate capacitances will still be calculated correctly).

Due to the effects of wiring capacitance, wiring resistance, and source/drain junction capacitance, the circuit characteristics before and after RC extraction can change dramatically, and, for example, a circuit that operated at 5 GHz before extraction may only run at 3 GHz in simulations after RC extraction. Finding this out after finishing layout would certainly turn our faces pale. We need some measure to prevent this from happening at the circuit design/simulation stage. I often either:

1. make **HDIF** larger than usual,
2. make AD/AS/PD/PS larger than usual

This, as shown in Fig. 4.10c, adds a dummy capacitance corresponding to the interconnect RC to the transistor source and drain. With the BSIM3 model, I overwrite the **HDIF** within the SPICE parameters. In cases such as if this is not allowed, or **HDIF** is not valid because the model is BSIM4, the netlist without AD/AS/PD/PS can be output, and your own AD/AS/PD/PS values can be calculated and overwritten. The "overwriting" will be very tedious to do by hand, so this will amount to writing a C program or perl script to do the job. In my case, when modifying **HDIF**, I use a value 2.5–3 times larger than the value determined by the process. When adding values for AD/AS/PD/PS, I let HDIF $= 5L$ (5 is a rule of thumb number and should be adjusted per process and design accordingly) and compute values for AD/AS/PD/PS based on the value of W and modify the netlist.

By taking these measures, the differences in characteristics before and after RC extraction can be minimized, and only minor adjustments will be required after RC extraction.

4.4 Typical Options

There are various options in RC extraction tools, and some of the options to be noted are discussed.

4.4.1 C Extraction and RC Extraction

When conducting interconnect extraction, this specifies whether to extract capacitances only or both resistances and capacitances (it is rare to extract only resistances). There is not a large difference in the time it takes to extract, but there is a large difference in the SPICE simulation run time with the extraction results. Therefore, if the effect of resistances is small, only capacitances should be extracted.

As a rule of thumb, in processes around 100 nm, wiring resistances are a few tens of m$\Omega/\square$, vias (wire to wire) are a few Ω per via, and contacts (wire to gate/source/drain) are a few tens of Ω per contact. Meanwhile, the ON resistances of transistors with regular (used for logic) transistor sizes are several hundred Ω to several kΩ as shown in Fig. 4.12, allowing the wiring resistance by comparison to be negligible. Conversely, for capacitances, wiring will have several tenths of fF per 1 μm, whereas transistor capacitances (the sum of gate capacitance and source/drain to substrate junction capacitances) are several fF per 1 μm of gate width, and so the wiring capacitance cannot be ignored. Therefore, only capacitances are normally

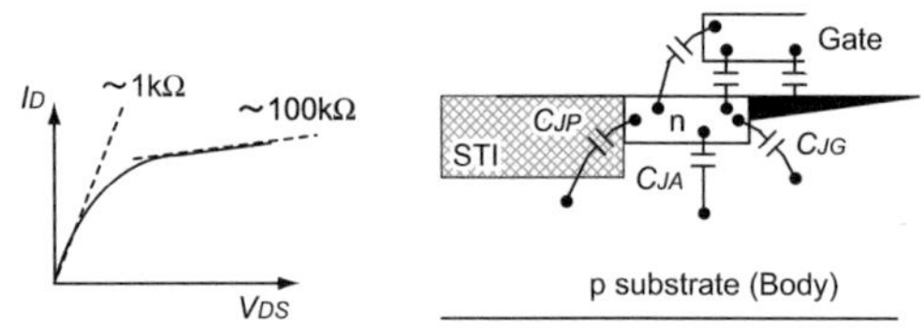

Fig. 4.12 Transistor resistance and capacitances

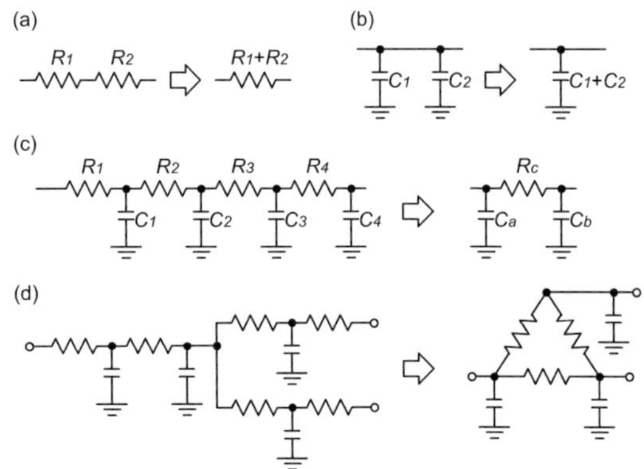

Fig. 4.13 RC compaction

extracted, and RC extraction is conducted for portions of interest such as long interconnect, clock, and large transistors.

4.4.2 Compaction

In capacitance extraction, regions must be divided into very fine pieces to apply to the aforementioned lookup table. Resistances are correspondingly divided and extracted. As a result, simulations become very time consuming due to the enormous RC network. RC extraction tools contain an RC compaction functionality, which reduces the number of RC elements while maintaining the accuracy. The transformations in Fig. 4.13a, b are obvious, but the tool can also transform the network in an RC ladder compaction as in Fig. 4.13c which maintains precision within a certain frequency range or from a Y network to a Δ network as in Fig. 4.13d, to reduce the number of components.

Obviously, compaction will change the characteristics of the circuit, and from the perspective of precision, things only get worse. The necessary precision and simulation time should be balanced to specify the compaction strength of the extraction tool.

4.4.2.1 Elmore Delay Model

To determine the delay (time constant) of an RC ladder, an approximation called the Elmore delay model is used. For example, the RC time constant τ_{Elmore} of Fig. 4.13c is

$$\tau_{\text{Elmore}} = C_1R_1 + C_2(R_1+R_2) + C_3(R_1+R_2+R_3) + C_4(R_1+R_2+R_3+R_4). \tag{4.21}$$

Here, if $R_1 = R_2 = R_3 = R_4 = R$ and $C_1 = C_2 = C_3 = C_4 = C$, the time constant can be approximated as

$$\tau_{\text{Elmore}} \approx 4R \cdot 4C/2. \tag{4.22}$$

The actual delay is $0.38 \times (4R \cdot 4C/2)$, so while the estimate is fairly accurate, we also see that the delay tends to be overestimated. To approximate this with the π model on the right of Fig. 4.13c, the resistance and capacitance values are set to $R_c = 4R$ and $C_a = C_b = 4C/2$.

There also exists a model called asymptotic waveform evaluation (AWE) as a model for compaction. This model determines the poles of the RC network, reduces the number of poles by merging nearby poles into an intermediate frequency, and simplifies the final RC network. You should at least remember the name.

4.4.3 *Dealing with Cross-Coupled Capacitances*

Interconnect capacitances are not always with respect to supply or ground and can exist between one wire and another, as shown in Fig. 4.14a, and in reality the interconnect RC extraction tool will output these kinds of netlists. Capacitances such

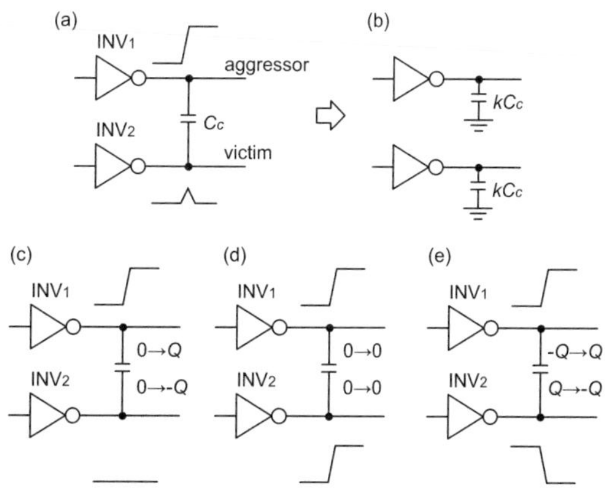

Fig. 4.14 Cross-coupled capacitances

as these which are not terminated with supply or ground but rather are connected between routing lines are called cross-coupled capacitances.

In this example, when the output of INV_1 changes, part of the signal travels through the capacitance C_c to the output voltage of INV_2, and spike noise is generated. This phenomenon is called cross talk, and in this example the output node of INV_1 is the source of noise or the aggressor, and the output node of INV_2 is the receiver of noise or the victim. In analog circuit design, the effects of cross talk noise, and not just wiring delay, must be taken into account.

Also, when simulating with Fast SPICE that we learned about in Sect. 2.2, this structure does not allow the partitioning of INV_1 and INV_2 which causes an increase in calculation time, so the cross-coupled capacitances are sometimes terminated to ground as in Fig. 4.14b such that the partitioning is possible. In this case, what is the effect of these capacitances on the delay? If we look at the delay of INV_1, the necessary charge to change the output of INV_1 changes with the operating condition of INV_2, as indicated in Fig. 4.14c–e. If the necessary charge for when the output of INV_2 is not changing is Q, as in Fig. 4.14c, then there is no need to store charge on C_c when INV_1 and INV_2 simultaneously change in the same direction, and a charge of $2Q$ must be moved when INV_1 and INV_2 simultaneously change in opposite directions, as in Fig. 4.14e. Therefore, to terminate the cross-coupled capacitance to ground with an equivalent capacitance as indicated in Fig. 4.14b, let $k = 1$ when the other signal does not move, $k = 0$ when the other signal moves simultaneously in the same direction, and $k = 2$ when the other signal moves in the opposite direction. In delay calculations, the design is safer when the wire delay is overestimated, so the conversion with $k = 2$ is common.

With extraction tools, all of the above should be taken into account to specify whether to keep the cross-coupled capacitance or to terminate them to ground with $k = 1$ or $k = 2$.

4.4.4 Dealing with Supply Lines

Supply lines span the entire chip and are an immense and complex routing. If RC extraction is conducted on supply lines, not only would a massive amount of RC elements be extracted, but also the Fast SPICE from Sect. 2.2 would not be able to partition the circuit leading to an enormous amount of simulation time. Thus, interconnect RC extraction is usually bypassed and simulations are run with the assumption of ideal power supplies.

Meanwhile, if the supply lines have resistance, then the supply voltage would drop due to the voltage drop from the transistor switching currents (IR drop). This can lead to an increase in delay and become the cause of circuit malfunction. Consequently, there exist special tools that determine the resistance distribution from the supply terminal to each transistor terminal using the RC extraction results of the supply lines and tools that determine the voltage drop distribution as shown in

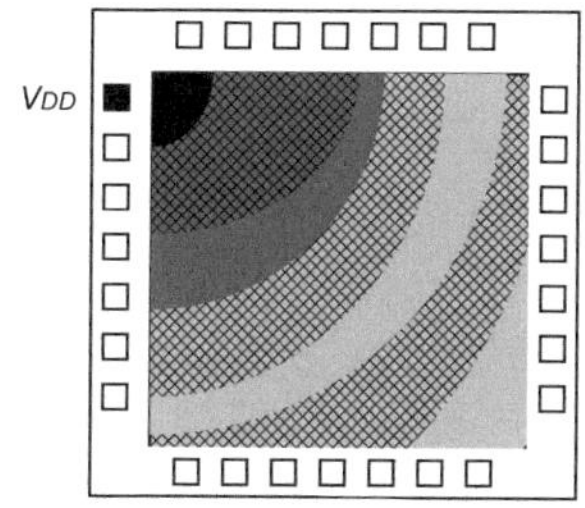

Fig. 4.15 Voltage drop distribution

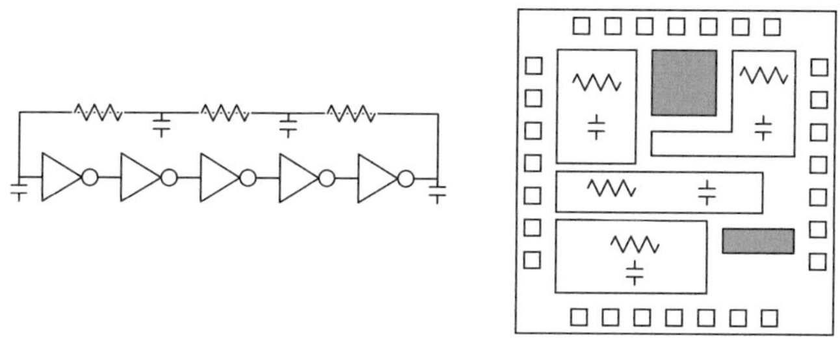

Fig. 4.16 Node specification and cell specification

Fig. 4.15 by estimating the current values from the switching activity of each block. These tools should be used as necessary.

Also, supply lines are, just as with CLK routing, special and important interconnect, and special attention is necessary during layout to avoid voltage drops, such as to increase the line widths or to create a mesh structure.

4.4.5 Node Specification and Cell Specification

It is possible to specify options such as to only extract the RC for a specified portion of the routing or to only extract (or not extract) the RC for some cells (SPICE SUBCKT units).

For example, as indicated in Fig. 4.16, an accurate RC extraction might only be necessary for a wire that may become long, and for others the parameters AD/AS/PD/PS could just be set more or less in excess. Also, the RC could be extracted only for important interconnect such as the clock tree or other signals that might be susceptible to noise. Additionally, RC extraction could be skipped for cells within the entire block that operate at a slower speed, or RC extraction could be conducted only for analog blocks and not digital blocks. In these ways, design and verification are conducted by balancing the overall accuracy and RC extraction with simulation time.

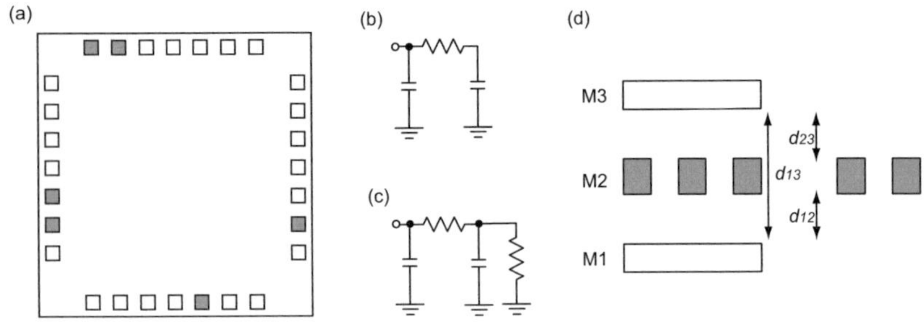

Fig. 4.17 Floating nodes and dummy fill

4.4.6 Dealing with Floating Nodes and Dummies

For example, in the pads in Fig. 4.17a, if there are unused pads, an independent capacitance with no DC path will be extracted. SPICE simulators cannot in general simulate circuits such as those in Fig. 4.17b because the node voltages become undefined. The simulation could be run as indicated in Fig. 4.17c by automatically inserting a large resistance to ground, but the simulations for these types of circuits have poor convergence, and the simulation will take a long time. To avoid these situations, it is recommended to specify the option to ignore floating nodes.

Also, as shown in Fig. 4.17d, if only M1 and M3 exist in the layout, M2 dummy is inserted to prevent unevenness on the surfaces due to CMP. It is a fairly difficult problem to choose running interconnect RC extraction between before dummy generation and after dummy generation. Naturally, running extraction against layout after dummy generation is closer to the actual situation, but when a large amount of dummy is inserted, there are problems such as:

- Interconnect RC extraction will take a long time
- After extraction, the number of nodes explodes and simulation will take a long time

Also, designers often do not look at (are not allowed to look at?) the layout after dummy generation to begin with, because dummy is inserted during mask fabrication. For these reasons, interconnect RC extraction is usually run against layout before dummy generation. For example, in the case of that in Fig. 4.17d, the distance between M1 and M3 with no dummy is d_{13}, but if the resistance of the dummy metal is ignored, then the equivalent distance between M1 and M3 with dummy is closer to $d_{12} + d_{23}$. Therefore, if RC extraction is run without taking dummy into consideration, a capacitance that is smaller than the actual value will be extracted. If the post-RC extraction simulation results do not match with actual measurements, these possibilities should be taken into consideration. Also, because a large number of floating nodes will be generated when running RC extraction with dummy fill, the option to ignore floating nodes is absolutely necessary.

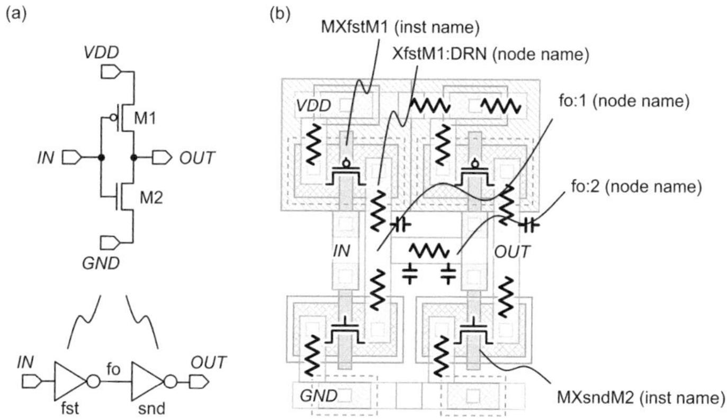

Fig. 4.18 XREF

4.4.7 XREF with LVS

Interconnect RC extraction will generate a netlist which contains transistors, resistors, and capacitors, but when the netlist including RCs is generated from the layout only, the portions except for the labeled terminal names will be assigned unintelligible names, such as M1, M2, NET1:0, NET1:1, NET2:0, NET2:1, and so on. If SPICE simulations are run directly from this, we would not be able to tell which node to choose for waveform observation.

Usually, interconnect RC extraction is run after LVS. Accordingly, node names in the netlist from interconnect extraction can be assigned based on the net names and instance names of the netlist that was referenced during LVS. In other words, when net names and instance names are given in the schematic editor as in Fig. 4.18, transistors will have names such as `MXfstM1` or `MXsndM1` (X indicates a SUBCKT instance in SPICE). Also, transistor terminals will have names such as `XfstM1:DRN` or `XfstM1:GATE`, and nodes other than transistor terminals will have names such as `f0:1` or `f0:2`. In this way, as long as meaningful names are given in the circuit schematic, the names in the post-extraction netlist will be easier to understand as well.

4.5 Reduction of Interconnect RC

Operating speeds become slower and power consumption increases due to interconnect RC. Here, we will examine methods to reduce interconnect RC.

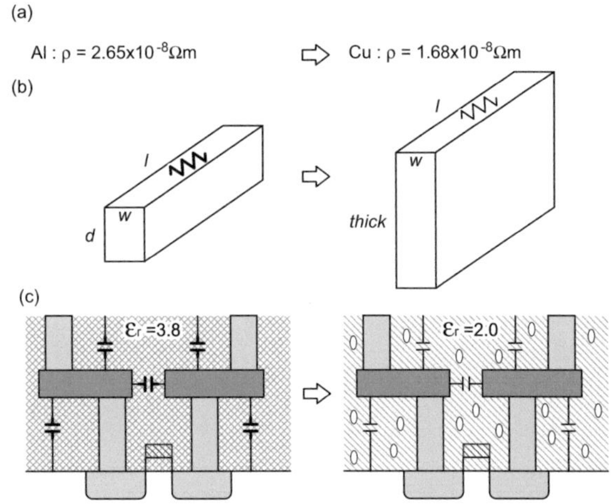

Fig. 4.19 (**a**) Al wiring and Cu wiring, (**b**) thick metal, (**c**) low-K materials

4.5.1 Process Technology

4.5.1.1 Al Wiring and Cu Wiring

The resistance values can be reduced by using low-resistivity materials in the wiring interconnect. Aluminum ($\rho = 2.65 \times 10^{-8}\,\Omega \cdot \mathrm{m}$) was formerly used, but recently the use of copper ($\rho = 1.68 \times 10^{-8}\,\Omega \cdot \mathrm{m}$) is more common (Fig. 4.19a).

4.5.1.2 Thick Metal

Resistance is inversely proportional to the cross-sectional area of the wire. The thickness of metal layers is usually several hundreds of nm, but by making this into a thickness of several μm, the cross-sectional area can be increased and the sheet resistance decreased (Fig. 4.19b).

4.5.1.3 Low-K

Parallel plate capacitance can be expressed as $C = \epsilon S/d$. The capacitance can be made smaller by using a material with low permittivity (low-K) for the insulating oxide film between routing lines. SiO_2 oxide film (dielectric constant approximately 3.8) was formerly used, but recently the permittivity is lowered by using organic polymers or introducing air pores into the insulating film (Fig. 4.19c).

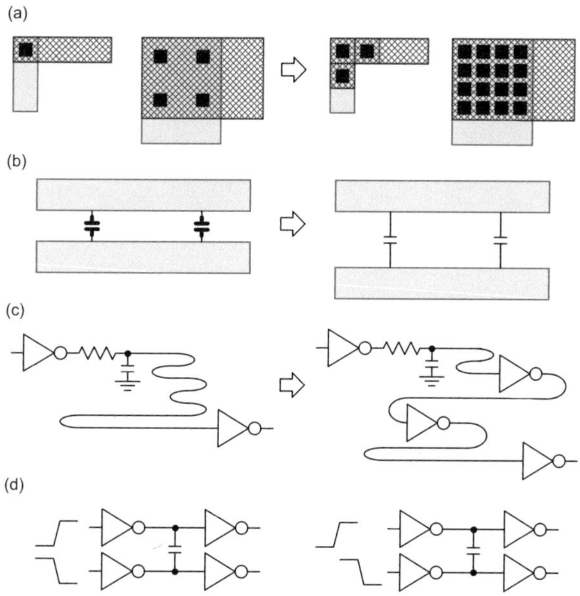

Fig. 4.20 (**a**) Multi-vias, (**b**) interconnect spacing, (**c**) repeaters, (**d**) simultaneous switching

4.5.2 Design Techniques

4.5.2.1 Multi-vias

Normally, the shapes of vias and contact holes are predetermined. Multi-vias and contacts are placed to reduce the resistance of vias and contacts. Supply lines in particular carry large currents, so the wires should be made wide and as many contacts should be placed as possible (Fig. 4.20a). Also by placing multi-vias, even in the case of a fabrication accident which causes a contact or via to fail to conduct, proper circuit operation can be achieved.

4.5.2.2 Interconnect Spacing

Capacitance can be lowered by increasing the spacing between a wire and a wire. This technique cannot be used if the routing is congested and wires must be placed at minimum distance intervals, but if there is area to spare, then the spacing should be made as wide as possible (Fig. 4.20b).

4.5.2.3 Repeaters

If a large resistance or capacitance exists due to a long wire, repeaters are inserted in the middle to prevent the deterioration of the rise and fall of the signal (Fig. 4.20c).

Also, the total delay is often lower when repeaters are used. This is not a method to reduce the interconnect RC per se but is often used as a design with interconnect RC taken into consideration.

4.5.2.4 Simultaneous Switching

Cross-coupled capacitances appear as an effectively doubled capacitance when simultaneous switching occurs in opposite directions, and the delay correspondingly increases. The circuit should be made with shifted timing of signals if possible so that simultaneous switching does not occur, which will prevent the increase of delay (Fig. 4.20d). This is also not a method to reduce the interconnect RC but is often used in the design of digital circuits as a design methodology that takes into consideration the interconnect RC.

Chapter 5
IO Buffers

We have designed the circuit that realizes the desired functionality, but how do we exchange signals with the outside of our chip? We simply connect lines on the circuit diagram, but it probably isn't so easy in reality...

5.1 Signal Path Between Chips

When chips are connected to each other and signals are exchanged, as shown in Fig. 5.1, the signal is output from the pad connected to the interior of the chip, which goes through the bonding wire, the lead frame of the package, and the transmission line on the board, which then goes through the other chip's lead frame, bonding wire, and pad, to finally arrive at the receive circuit as input.

These channels have impedances which are a complicated combination of resistances, capacitances, and inductances. Therefore, to transmit and receive signals cleanly through these kinds of channels, it is necessary to estimate the impedances of these channels and design corresponding input-output circuits.

5.1.1 Pads

In a chip layout as in Fig. 5.2a, rectangular shapes in rows on the chip periphery can often be observed. These are called the pads (or IO pads). In recent processes, the pad pitch is about 60 μm. Also, because these take up a large area, they have a parasitic capacitance with the substrate surface of several hundred fF $\sim$ 1 pF.

The chip surface is covered by a layer called passivation film so that dust and moisture do not enter the inner parts of the chip. Polyimide is often used as the material for the passivation layer. Here, the passivation has a window of opening for

T. Nakura, *Essential Knowledge for Transistor-Level LSI Circuit Design*,
DOI 10.1007/978-981-10-0424-7_5

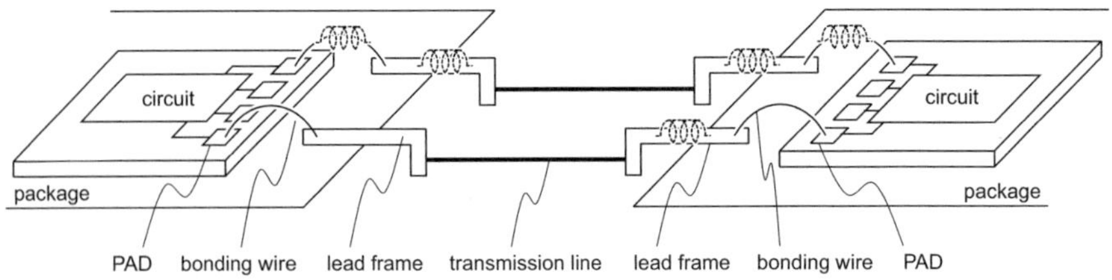

Fig. 5.1 Signal path between chips

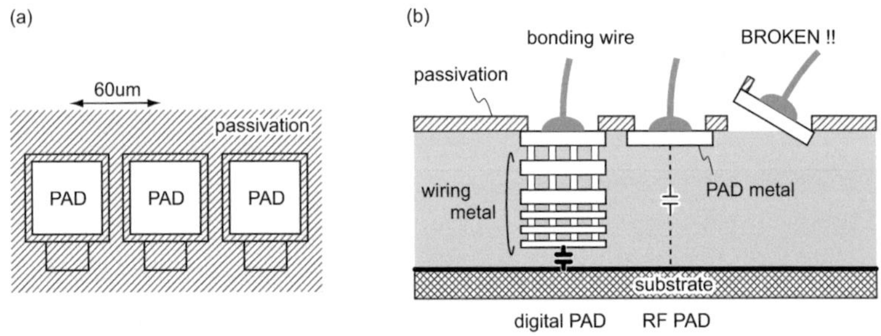

Fig. 5.2 (**a**) Pad layout, (**b**) pad cross section

the pads, and signals are transmitted and received between the inside and outside of the chip through this window.

The cross section of pads is shown in Fig. 5.2b. The uppermost layer of metal is called the pad metal and is used as pads. Underneath the pads, all routing layers connected by vias could be placed (digital pads), or nothing could be placed at all (RF pads). RF pads can be used for the input and output of fast signals, because RF pads have smaller parasitic capacitances to the substrate. However, as indicated in Fig. 5.2b, RF pads are more vulnerable to damage during bonding, so digital pads that have higher structural integrity are also used often. Although the parasitic capacitance will increase, this is better than malfunctions.

5.1.2 Packages and Bonding Wires

Packages come in many shapes and forms, such as quad flat package (QFP), small outline package (SOP), ball grid array (BGA), and so on. As for the material, there are ceramic packages that use ceramics and plastic packages that use plastic. In general ceramic packages have multiple functionalities and options, such as built-in transmission lines, added capacitance to the package internals, and structures with removable covers. However, these can become very expensive, and a single package can cost around 100USD. Plastic packages on the other hand do not have such capabilities but are priced at several cents a piece.

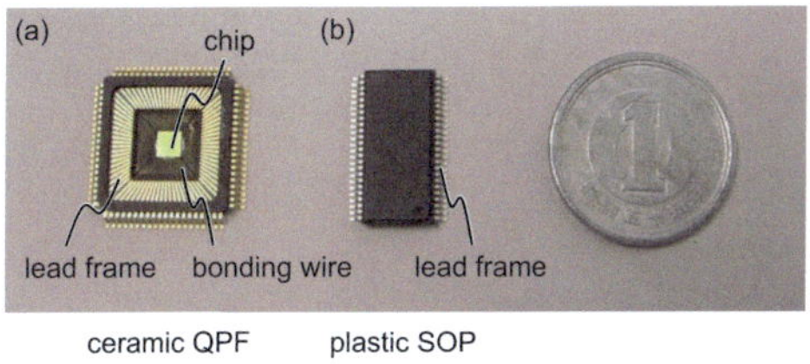

Fig. 5.3 Photograph of packages

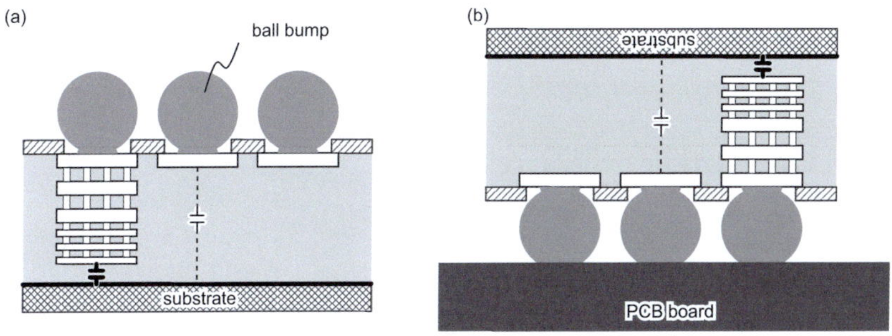

Fig. 5.4 Flip chip bonding, (**a**) forming bumps, (**b**) bonding

A photograph of a ceramic QFP and plastic SOP that were at hand is shown in Fig. 5.3. The cover of the ceramic package is opened so that the bonding wires and lead frames are also visible. Gold wires with diameters of tens of μm are typically used for the bonding wires.

Now, it can be readily imagined that the lead frame and bonding wires have some impedance. As a rule of thumb, it is good to remember "1 nH per 1 mm." Of course there are also resistance and capacitance components, but these are negligible compared to the inductance. In Fig. 5.3, the inductance is several nH. Inductance needs to be suppressed, because otherwise the waveforms of fast signals can become distorted or large supply noise can be generated due to internal switching. Also, package sizes are large relative to chip sizes, and this becomes a hindrance to miniaturization of instruments. To solve these issues, strategies such as chip-size package (CSP) or flip chip bonding, which does not use packages (Fig. 5.4), are sometimes used.

However, because handling the chips is difficult with flip chip bonding, in recent years, multi-chip packages as illustrated in Fig. 5.5 are also being used. By doing so, a low impedance connection between chips exchanging fast signals can be made possible by placing them in the same package, while achieving miniaturization as well.

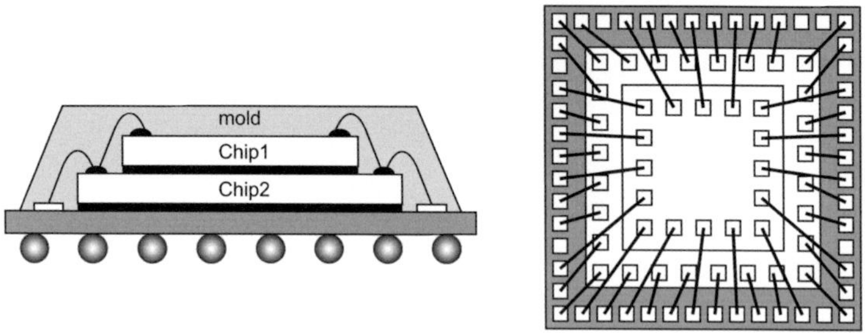

Fig. 5.5 Multi-chip package

5.1.3 Transmission Lines

5.1.3.1 Characteristic Impedance

When a voltage V_0 is applied to resistances R_s and R_L, which are connected by a "long" wire, as shown in Fig. 5.6a, what happens? Would the current become $V_0/(R_s + R_L)$ from the instant the voltage rises? When and how would the current know that a resistance R_L exists at the end of this "long" wire?

If the resistance of the wire is ignored, all wires (regardless of whether the wire is "long" or not) contain inductance and capacitance as indicated in Fig. 5.6b. If L and C are the inductance and capacitance per unit length,

$$Z_0 = \sqrt{\frac{L}{C}} \tag{5.1}$$

is called the characteristic impedance. The current, which does not know what is connected at the end of the long wire, will enter the wire believing that the characteristic impedance Z_0 is connected at the end, as indicated in Fig. 5.6c. That is, a current of $I = V_0/(R_s + Z_0)$ initially starts to flow. This current reaches the resistance R_L at the speed of light c. In spite of the current's belief that Z_0 was connected, it will discover that in fact R_L is connected, and . . . then what?

5.1.3.2 Termination and Reflection

Various phenomena occur by adjusting the impedance connected at the end of the "long" wire. This is called "termination."

The initial current that flows into the wire I_{1f} and initial voltage V_{1f} are $I_{1f} = \frac{V_0}{R_s+Z_0}$ and $V_{1f} = \frac{Z_0}{R_s+Z_0}V_0$, respectively. When the wire is terminated by Z_0 ($R_L = Z_0$), the current that came along the wire believing that there is a Z_0 at the end actually finds the Z_0 and flows directly into G_{ND}. This is called matched termination or impedance matching.

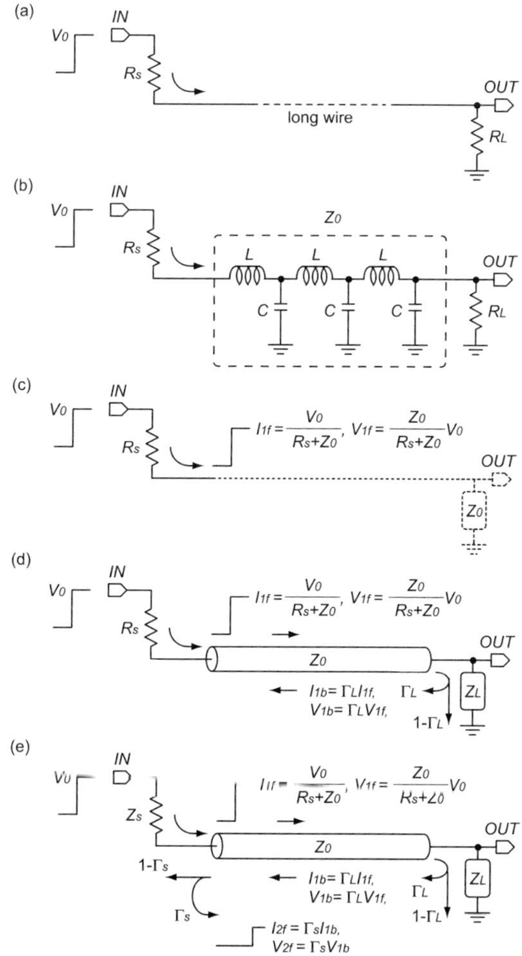

Fig. 5.6 Transmission line characteristic impedance, termination, and reflection

On the other hand, if there is not a Z_0 but a Z_L that is connected, then part of the current will "reflect" and come back, as indicated in Fig. 5.6d. If the reflectance is Γ_L, then the reflected current and voltage I_{1b} and V_{1b} are $I_{1b} = \Gamma_L I_{1f}$ (left is the positive direction) and $V_{1b} = \Gamma_L V_{1f}$, respectively. The current that does not reflect back but enters Z_L is $I_{1L} = (1 - \Gamma_L)I_{1f}$. Here, the voltage at the right end is $V_{1L} = V_{1f} + V_{1b}$, and this is equal to the voltage that is generated on the termination Z_L. That is,

$$V_{1f} + \Gamma_L V_{1f} = (1 - \Gamma_L)I_{1f}Z_L \tag{5.2}$$

and the reflectance Γ_L can be derived from this equation and $V_{1f} = I_{1f}Z_0$ as

$$\Gamma_L = \frac{Z_L - Z_0}{Z_L + Z_0}. \tag{5.3}$$

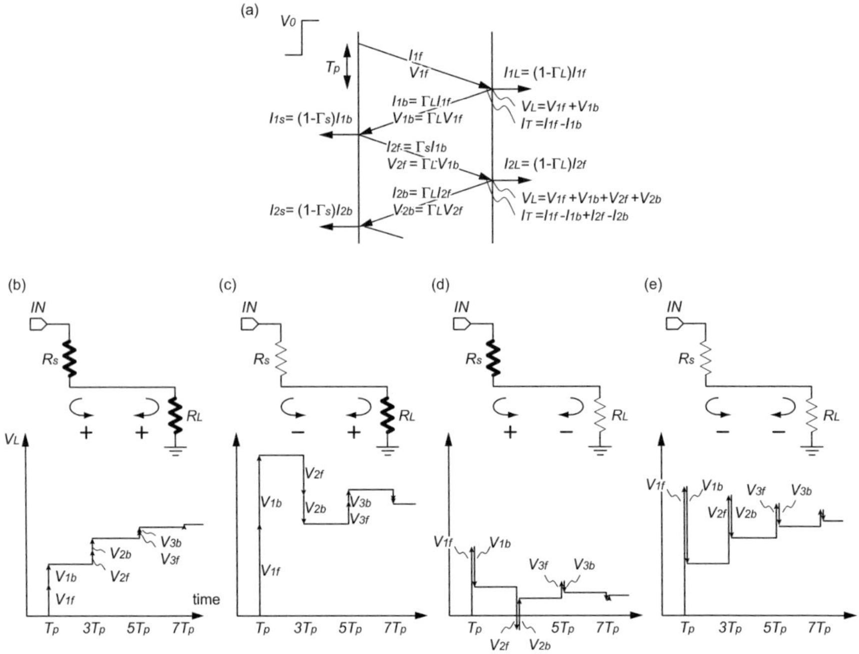

Fig. 5.7 Classification of reflections based on termination impedance magnitudes

The current that reflected and came back is reflected again as shown in Fig. 5.6e:

$$\Gamma_s = \frac{Z_s - Z_0}{Z_s + Z_0} \tag{5.4}$$

and a portion flows to the right. These reflections are repeated until the currents and voltages settle to their final values (the equations converge when $|\Gamma_s \Gamma_L| < 1$).

Here, for example, in Fig. 5.6e, the current going to the right is $I_{1f} - I_{1b} + I_{2f}$, and the voltage is $V_{1f} + V_{1b} + V_{2f}$. Be aware that the current changes signs depending on whether it is traveling right or left, whereas the voltage has no sign.

The reflecting signals are schematically indicated in Fig. 5.7a. The traveling signals are indicated on the horizontal axis, and the vertical axis indicates time. At time T_p, the signal arrives at the rightmost point (receiving end), and the voltage at that time is $V_{1f} + V_{1b}$. The reflected wave V_{1b} is again reflected at the leftmost point (transmitting end), and at time $3T_p$, the re-reflected wave arrives at the receiving end. The voltage at that time is $V_{1f} + V_{1b} + V_{2f} + V_{2b}$. According to Eq. (5.3), the reflection coefficient is positive when the termination impedance is greater than the characteristic impedance ($Z_L - Z_0 > 0$) and negative when the termination impedance is less than the characteristic impedance ($Z_L - Z_0 < 0$). Therefore, there are four possible combinations of the impedances of the transmitting and receiving ends being greater than or less than the characteristic impedance. For each situation, the voltage waveform at the receiving end is shown in Fig. 5.7b–e. The voltage can asymptotically approach the final voltage monotonically or by oscillating.

As an extreme example, when $Z_s = Z_0$ (matched termination) and $Z_L = \infty$ (open termination), the situation is in between Fig. 5.7b, c, and $I_{1f} = V_0/(2Z_0)$, $V_{1f} = V_0/2$, $\Gamma_L = 1$, $I_{1b} = \Gamma_L I_{1f} = I_{1f}$, $V_{1b} = \Gamma_L V_{1f} = V_{1f}$, $\Gamma_s = 0$, $I_{2f} = 0$, and $V_{2f} = 0$. That is, the voltage at the receiving end is zero until the signal arrives, and at time T_p becomes $V_0/2$, but the arrival and reflection occur simultaneously and the voltage becomes V_0 ($V_{1f} = V_{1b} = V_0/2$, $V_{1f} + V_{1b} = V_0$). The current traveling right is zero until the signal arrives, and the moment the signal arrives, it becomes I_{1f}, but the arrival and reflection occur simultaneously and the current becomes zero ($I_{1f} = I_{1b}$, $I_{1f} - I_{1b} = 0$). The voltage at the transmitting end is $V_0/2$ until the reflected wave arrives and becomes V_0 when the reflected wave does arrive. The current traveling right at the transmitting end is I_{1f} until the reflected wave arrives and zero after the reflected wave does arrive. Another reflection does not occur ($\Gamma_s = 0$).

5.1.3.3 Lumped Parameter Circuits and Distributed Parameter Circuits

Strictly speaking, all wires are transmission lines, and as shown in Fig. 5.8a, even when an inverter drives the next inverter stage inside an LSI, a steady state is reached after repeating such reflections within the wire. However, reflections are not observed because the wire is short and the period of reflection is short. Thus, as shown in Fig. 5.8b, there is no problem approximating this wire as a resistance and capacitance, and this is called a lumped parameter circuit. Meanwhile, as in Fig. 5.6, circuits with elements distributed throughout are called distributed parameter circuits. As a rule of thumb, when the wire length is less than 1/6 of the wavelength, the circuit can be treated as a lumped parameter circuit, whereas if the wire length is greater than 1/6, then the wire must be treated as a transmission line, which is a distributed parameter circuit. For example, with 10 GHz signals, the wavelength in vacuum is 3 cm, and 1/6 of that is 5 mm. Thus, for LSI operating at 10 GHz, wires longer than 5 mm (50 cm if 100 MHz) must be treated as transmission lines.

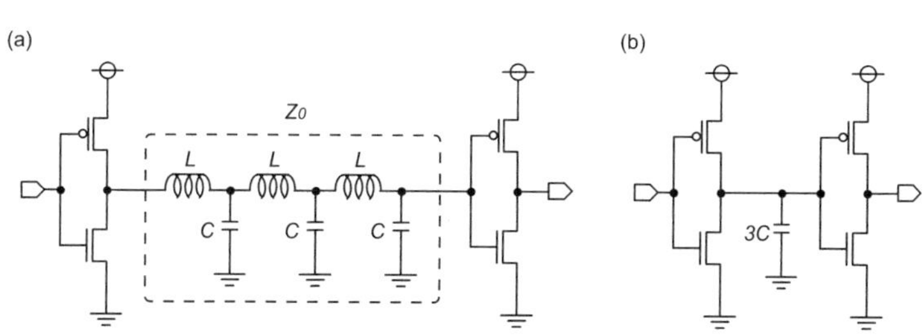

Fig. 5.8 Distributed and lumped parameter circuits

5.1.3.4 50 Ω

50 Ω is most often used as the characteristic impedance. Some people say that this standard came about as a round number that is between 30 Ω that is easy to use from a power transfer perspective and 75 Ω which is easy to use from the signal loss perspective.

5.1.4 Termination Methods

As shown in Fig. 5.9, there are many methods for termination. For the following, the transmitting and receiving circuits are both CMOS inverters, and $Z_0 = 50\,\Omega$:

(a) This is the situation with no termination and is used when the signal line is short relative to the transmission rate, or in other words the wire can be treated as a lumped parameter circuit. The signal levels are at full swing between V_{DD} and G_{ND}.
(b) This is called parallel termination and is the simplest form of termination. When the input impedance of the receiving circuit is infinite, the termination is a resistance of $R_T = Z_0$ (e.g., when the input of the receiving circuit is connected to the gate of a MOS, the gate leakage current is negligible, and the impedance of the gate capacitance is much greater than $50\,\Omega$: $|1/j\omega C| \gg |Z_0|$). Inputs of measurement equipment such as oscilloscopes mostly employ this type of structure, and at the research and development level, this is used often from the compatibility with laboratory measurements. However, there are drawbacks, such as a large power consumption, dependency of the power consumption on the signal level H/L ratio (the power consumption is higher when the percentage of H is higher), and an imbalance in the sense that the L level goes all the way

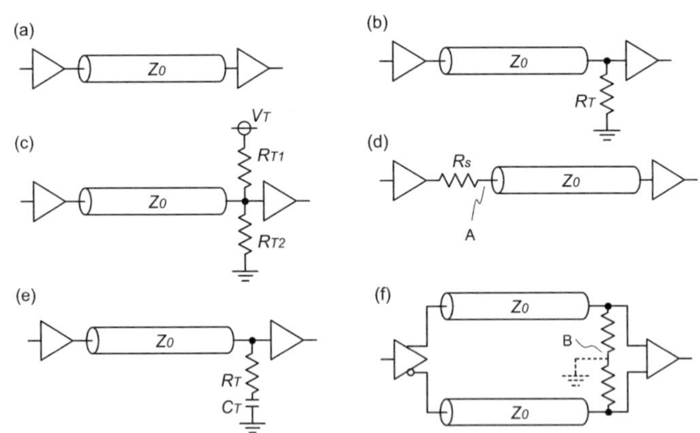

Fig. 5.9 (**a**) No termination, (**b**) parallel termination, (**c**) Thevenin termination, (**d**) series termination, (**e**) AC termination, (**f**) complementary parallel termination

down to G_{ND}, but the H level only goes up to $R_T V_{DD}/(R_o + R_T)$. Here, R_o is the output impedance of the transmitting circuit.

(c) This is called the Thevenin termination and is designed so that $R_{T1}//R_{T2} = 50$. Also, R_{T1}, R_{T2}, V_t can be adjusted according to the output impedance of the transmitting circuit to adjust the output signal level. In addition, the power consumption does not change while transmitting H or L. However, there are problems such as the increase in the number of elements for termination, the necessity of a V_t voltage source, and a power consumption even when the output is floating.

(d) This is called the series termination, and R_s is determined as $R_o + R_s = 50$. Here, the voltage at A is the instant after the signal output is $V_{DD}/2$. When this signal arrives at the receiving end, the signal reflects with reflectivity $\Gamma_L = 1$ (assuming the receiving end is open), so the voltage at the receiving end becomes V_{DD}. When the reflected voltage returns to the transmitting end, no further reflections occur and the signal is absorbed by the transmitting circuit because this end is terminated with 50 Ω. This method has a low power consumption and is often used in 1:1 communication, but cannot be used in 1:N communication.

(e) This is called AC termination and is used for fast signal transmission where $1/j\omega C_T \ll 1$. This has a low power consumption because there is no DC current flow, but for slower signals, C_L will appear as a load capacitance which may slow down the signals. Also, this can only be used if the H and L are equally likely to occur, and there are no long continuous streams of 1's or 0's. In addition, there is an increase in cost and difficulty in implementation due to the need to separately use a C_T.

(f) This is called the complementary parallel termination. When the two lines always transmit a 0/1 or 1/0 complementary signal, then point B at the receiving end becomes a virtual ground and the voltage does not fluctuate. Therefore, when each transmission line is terminated by 50 Ω to point B, the circuit becomes the parallel termination of (b), but because point B is virtual ground, it simply suffices to terminate between the two lines with a resistance of 100 Ω. This is used widely in the transmission of complementary signals.

5.1.5 Voltage Levels

When transmitting signals, several voltage levels have been standardized, such as H:V_{DD}/L:G_{ND} or H:1.425 V/L:1.075 V. Here, things can get difficult if the transmitting side and receiving side use different processes and the supply voltages are different. For example, when a 65 nm 1.2 V chip and a 0.35 μm 3.3 V chip communicate with each other, the 65 nm chip must use a 3.3 V high-voltage transistor just for the IO. Some examples of the standards are low-voltage differential swing (LVDS), pseudo emitter-coupled logic (PECL), low-voltage PECL (LVPECL), and current mode logic (CML), and these voltage levels are summarized in Table 5.1. High-speed signal transmission standards are descendants of the previously popular

Table 5.1 IO voltage levels

	LVDS	PECL(5 V)	LVPECL(3.3 V)	CML	CMOS
TX V_H	1.425 V	4.0 V	2.3 V	V_{DD}	V_{DD}
TX V_L	1.075 V	3.2 V	1.6 V	$V_{DD} - 0.8$ V	G_{ND}

emitter-coupled logic (ECL: H levels at −0.9 V, L levels at −1.7 V) for bipolar transistors with supply voltages of −5.2 V, and PECL can be thought of as a replacement by CMOS.

5.2 ESD

We all have the experience of the shock when touching a doorknob while wearing a sweater during a dry winter. Here, it is said that a voltage of thousands of volts exists momentarily. This is called electrostatic discharge (ESD). What would happen if this discharge was not with a doorknob but with an LSI chip? Certainly, the chip would be destroyed.

5.2.1 ESD Models

For this ESD, there are several standard discharge models, as shown in Fig. 5.10.

(a) This is a model called the human body model (HBM) and is a model where a person with electrical charge touches a chip. The experiment is conducted by applying a voltage of several hundred to several thousand volts, with $C_D = 100$ pF and $R_D = 1.5$ kΩ. The high voltage is applied over a period of several hundred ns.
(b) This model is called the machine model (MM) and is a model where a machine with electrical charge touches a chip. A charged $C_D = 200$ pF capacitor is brought into contact with the chip without a resistance. A voltage of 100∼400 V is applied. The high voltage is applied across several ns.
(c) This model is called the charged-device model (CDM) and is a model where the chip with electrical charge touches a conductor and discharges. This can also be thought of as a variant of MM. A voltage of several hundred volts is applied, across several ns.

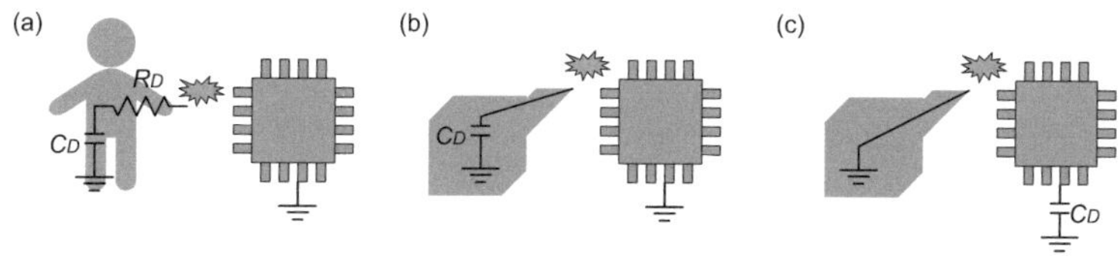

Fig. 5.10 (**a**) HBM, (**b**) MM, (**c**) CDM

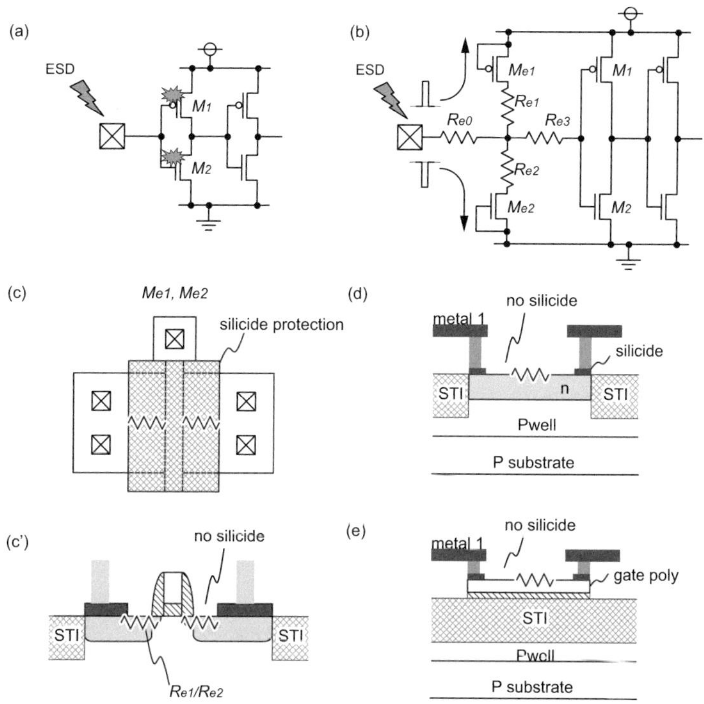

Fig. 5.11 (**a**) No protection circuitry, (**b**) a general form of ESD protection circuitry, (**c**) S/D resistance layout, (**c′**) S/D resistance layout cross section, (**d**) well resistance, (**e**) gate poly resistance

5.2.2 ESD Protection Circuits

To prevent chip destruction due to ESD, it is a basic practice to use wrist bands connected to G_{ND} so that ESD does not occur. However, ESD is something that occurs in unexpected situations, such as during chip bonding or transportation. Thus, it is necessary to embed ESD protection circuitry within the IO buffers, which are the points of contact between the chip and the outside world, so that the chip does not break even when ESD occurs.

As indicated in Fig. 5.11a, the gate oxide film of transistors M_1 and M_2 will be destroyed when ESD is applied. A representative ESD protection circuit is indicated in Fig. 5.11b. By allowing the injected ESD pulse to pass through the resistance R_{e0}, the voltage is reduced. At the same time, by allowing the charge to escape to V_{DD} and G_{ND} through diode-connected transistors M_{e1} and M_{e2}, the application of a high voltage onto transistors M_1 and M_2 is prevented. That is, when a high positive voltage is applied, the gate and (originally) source terminals of M_{e1} that are connected to V_{DD} become a lower voltage relative to the PAD or (originally) drain terminal. Thus, M_{e1} turns on, the ESD pulse flows to V_{DD}, and no load is applied to M_1 and M_2. Similarly, when a high negative voltage is applied, the gate and (originally) source terminals of M_{e2} that are connected to G_{ND} become a higher voltage relative to the PAD or (originally) drain terminal. Thus, M_{e2} turns on, the ESD pulse flows to G_{ND} (the current flows from G_{ND} to PAD), and again no load is applied to M_1 and M_2. Here, resistances R_{e1} and R_{e2} are often inserted so that the ESD protection transistors M_{e1} and M_{e2} themselves do not break. As shown in Fig. 5.11c, c′, silicide protection for preventing the accumulation of low-resistance silicide is applied in the S/D regions of M_{e1} and M_{e2} to realize resistances R_{e1} and R_{e2}. Well resistances without silicide and gate poly resistances without silicide are often used for resistances R_{e0} and R_{e3}, as shown in Fig. 5.11d, e.

5.2.3 *Miscellaneous Topics Regarding ESD Protection Circuitry*

5.2.3.1 Input Buffers and Output Buffers

In general, gate oxide films are more susceptible to breaking than PN junctions. Therefore, as shown in Fig. 5.12a, b, input buffers break more easily than output buffers. However, when a termination resistance R_t is used as in Fig. 5.12c, the transistor is less likely to break due to the ESD discharge flowing through R_t.

5.2.3.2 Speed Degradation

When resistances $R_{e0} \sim R_{e3}$ are inserted as in Fig. 5.11b, the transmission speed of IO buffers degrades. Sometimes only M_{e1} and M_{e2} are used without resistances as

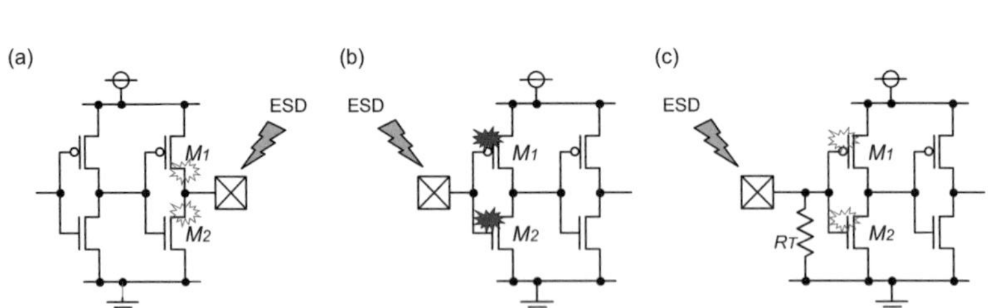

Fig. 5.12 (**a**) Output buffer, (**b**) input buffer, (**c**) input buffer with termination

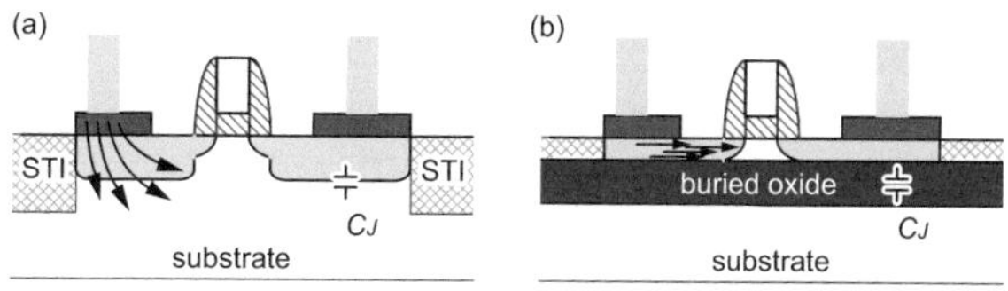

Fig. 5.13 (**a**) ESD discharge path for regular MOS, (**b**) ESD discharge path for SOI

the ESD protection circuit to prevent this degradation of transmission speed, but in this case, the ESD tolerance degrades. Also, while IO buffers themselves have load capacitances (the gate capacitances of M_1 and M_2 in Fig. 5.11b), the addition of ESD protection circuitry increases the load capacitance (drain capacitances of M_{e1} and M_{e2} in Fig. 5.11b) and invites further speed degradation. From the perspective of ESD protection, the gate widths W of M_{e1} and M_{e2} should be as large as possible. However, this leads to speed degradation due to the increased capacitance. Therefore, there is a trade-off relationship between ESD tolerance and transmission speed. Care is especially needed with the input and output terminals of high-speed, small analog voltages such as the LNA input terminal in an RF circuit.

5.2.3.3 ESD Tolerance in SOI

Silicon on insulator (SOI) has a structure as shown in Fig. 5.13b and is used in ultrahigh-speed LSI. This is because relative to a regular MOS shown in Fig. 5.13a, the source and drain junction capacitances C_Js are smaller and thus a low power consumption and fast operation are possible. However, the crystalline structure of the thin silicon layer on top of the buried oxide layer is problematic, and the gate oxide layer tends to break easily. Additionally, a large instantaneous current flows through the transistors for ESD protection (M_{e1} and M_{e2} of Fig. 5.11b). This current density is lower in regular CMOS because the ESD discharge disperses into the substrate, as indicated in Fig. 5.13a. However, the current density in SOI is higher because the path for ESD discharge is narrower as shown in Fig. 5.13b, and damage due to heat can occur. In this way, while SOI is a promising next-generation transistor structure, the lack of ESD tolerance is a weakness.

5.3 Types of IO Buffers and Their Layout

Examples of IO buffers include buffers for internal voltage supply, internal ground, IO supply, IO ground, bias voltages, low-speed digital signal outputs, low-speed digital signal inputs, high-speed digital signal outputs, high-speed digital signal inputs, analog signal outputs, and analog signal inputs. Also, termination resistances are embedded in the signal input and output buffers as necessary, along with ESD protection circuitry. As for the supply lines, various supply lines can be cross-connected with ESD protection circuitry in some cases.

5.3.1 IO Buffer Examples

Some representative examples of IO buffers are shown in Fig. 5.14.

(a) This is for the input and output of low-speed digital signals, where ESD protection circuitry is connected to CMOS inverters. The signals are full swing between V_{DD} and G_{ND} levels. The signals will reflect because there is no termination, and therefore these can only be used for low-speed transmission for which the signal integrity is not affected by the reflections that occur when the signal rises and falls. Also, while these are "low speed," fairly large transistors are required to be able to drive the several pF of pad capacitance.

(b) This is for the input and output of low-speed digital signals. When the Z terminal is pulled high (H), M_{z1} and M_{z2} both turn off, and the output becomes high impedance (High Z). There are three possible output states of high (H), low (L), and high impedance (High-Z), and therefore this is called a tristate buffer. When receiving signals from tristate buffers, a large resistance R_{PD} is often placed in the input buffer to prevent instability of the internal states due to receiving a High-Z signal and the input becoming indeterminate. By doing so, when the input is High-Z, the input buffer will pull its own input down to the G_{ND} level through R_{PD}. This resistance R_{PD} is called the pull-down resistance. To make sure that the V_{DD} level is reached when the signal is H, a resistance value significantly larger than the output impedance of the output buffer must be used for R_{PD}. Also, if this resistance is connected to the supply side, it becomes a pull-up resistance, and when the input is High-Z, the voltage is pulled up to V_{DD}.

(c) This is for the input and output of high-speed digital signals. ESD protection circuitry and a termination resistance R_T are connected to CMOS inverters. A signal of H will not reach V_{DD}, and the voltage level will be determined by the resistive divider between R_T and the on-resistance of M_1. A signal of L will go down all the way to G_{ND}. Multiple reflections can be prevented by making the on-resistances of M_1 and M_2 approach Z_0. A constant, steady-state current flows when outputting H.

(d) This is called open drain and is a different form for the input and output of high-speed digital signals. For a signal of H, V_{DD} is reached by turning M_2 off. The level for a signal of L is determined by the resistive divider of R_T and the on-resistance of M_2. Multiple reflections can be prevented by making the on-resistance of M_2 approach Z_0. A constant, steady-state current flows when outputting L.

(e) A typical example for the input and output of analog signals. By outputting with an NMOS source follower, the impedance seen from the internal pin A is infinite (the gate capacitance of M_2). At the same time, the analog voltage can be output from the internal pin with an AC gain of 1 (0 dB). The usable voltage range depends on the value of R_T and the size of M_2. Also, care is necessary when using this, because the characteristics change depending on whether the input pin B is, for example, connected to a gate of a MOS transistor and thus

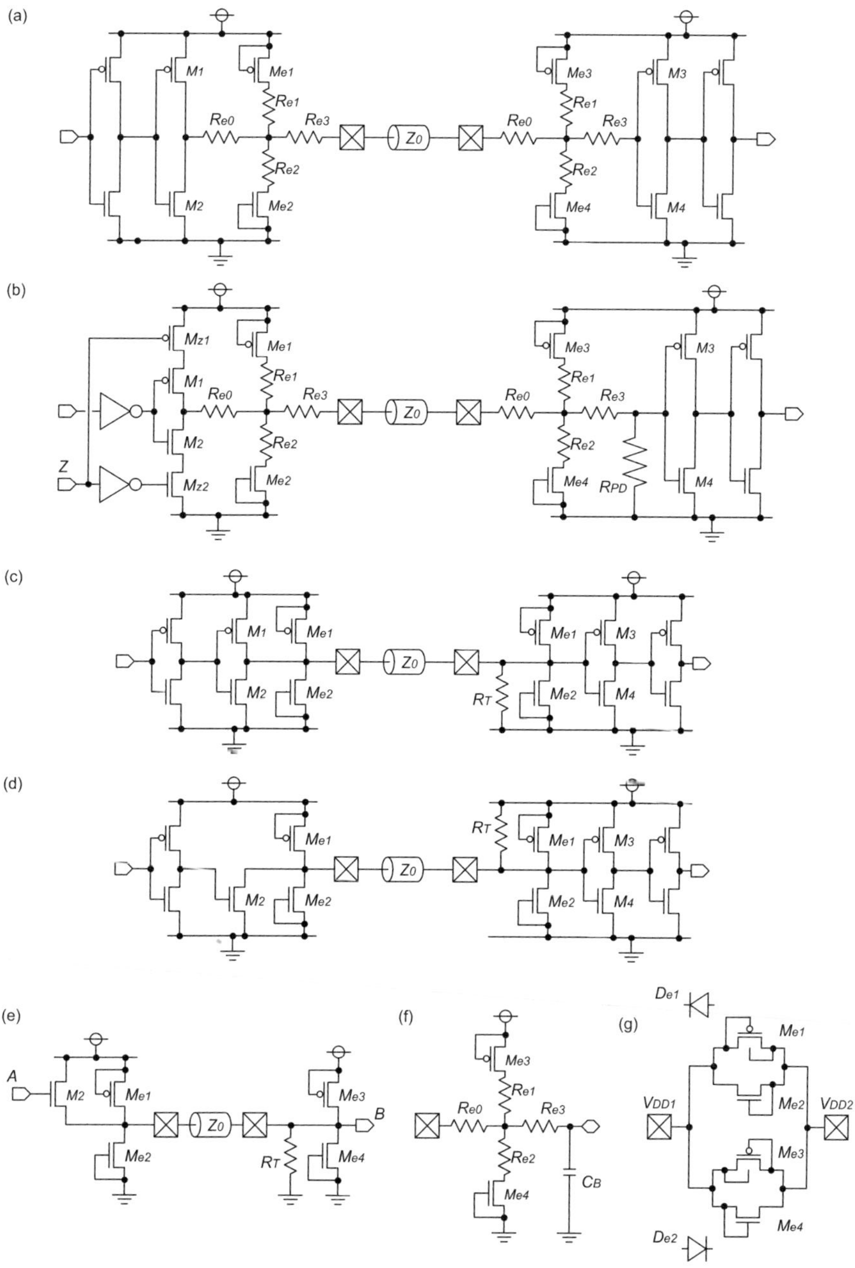

Fig. 5.14 Types of IO buffers: (**a**) low-speed digital; (**b**) tristate output and pull-down input; (**c**) high-speed digital; (**d**) high-speed digital, open drain; (**e**) analog, source follower; (**f**) bias voltage; (**g**) connecting various supplies to the same voltage

will not carry any current or is connected to the drain and there will be current flow.

(f) This is a buffer used for the input and output of bias voltages. DC current does not flow, and a voltage is assumed to be input and output. If DC current will flow, $R_{e0} \sim R_{e3}$ must be removed. In addition, the R and C can also be removed to input and output analog signals.

(g) This is used to supply the same voltage to the same chip for different means, for example, a 1.8 V supply for digital circuits and a 1.8 V supply for analog circuits. Bidirectional diodes are connected as a countermeasure against ESD, so that, for example, if a high voltage is applied to V_{DD1}, M_{e3} and M_{e4} turn on and charge can be released to V_{DD2}. Also, when different voltages are applied to the same chip, for example, a 3.3 V IO supply to V_{DD1} and a 1.8 V internal circuit supply to V_{DD2}, M_{e3} and M_{e4} are removed and only the reverse-biased diode is used.

5.3.2 Supply Rings

In general, the supply for the internal circuits and the supply for IO are separated, because IO consumes a large current and generates noise. Pads are often laid out in the periphery of the chip, and it is common to give the pads their supply and ground by creating rings for IO supply and IO ground as shown in Fig. 5.15. While there exist various types of IO, they are easy to use when their layouts are the same size so that the IO supply, IO ground, and pads are correctly connected simply by replacing the cell. Of course, ESD protection circuitry and termination resistances are also included within the IO cell.

5.4 Determining Pin Placement

As mentioned in Sect. 5.1.2, IO on the chip are connected to the routing on the board through the pad, a bonding wire, and the package lead frame. When determining the pin placement, the electrical characteristics of these must also be taken into account.

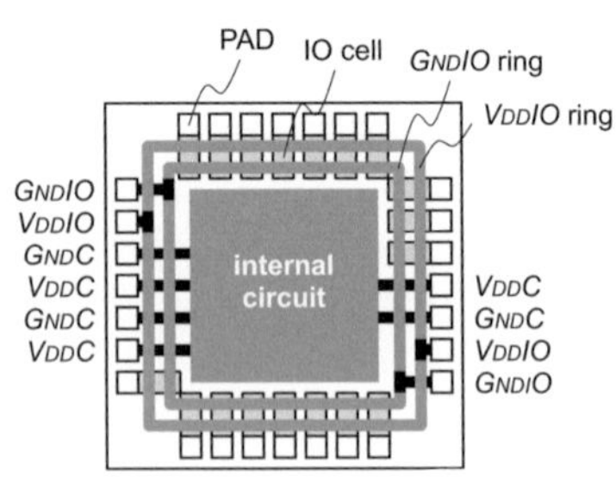

Fig. 5.15 Supply ring

5.4.1 Supply Pin

In general, a low impedance is desirable for supply lines. To reduce the bonding wire resistance and contact resistance of the supply lines, multiple supply pins, rather than a single pin, are used. Here, to prevent the voltage from dropping due to the internal supply line resistance of the chip, it is desirable for all circuits to be equidistant from a supply pin. Therefore, it is good to place a second supply pin on the opposite side of the chip, as shown in Fig. 5.15.

The inductance of the supply lines is also problematic. As we know, inductance is determined by the area of the loop of current. Thus, rather than placing the chip V_{DD} and G_{ND} far apart as in Fig. 5.16i, the inductance can be made smaller by placing them next to each other as in (ii). Also, when using multiple pins, rather than placing the pins as VVGG as in (iii), a placement of VGVG as in (iv) will make the inductance smaller.

5.4.2 Shielding

Let's say S_0 is a high-speed signal or a noise-sensitive analog signal, and cross-talk noise through parasitic capacitance could be injected by adjacently placed signal lines as in (v). Therefore, the signal should be placed between ground lines if possible as in (vi), to prevent cross-talk noise. Also, because the loop area of the signal and ground lines is reduced by doing this, not only will the inductance be reduced, but also the termination to ground will be more effective.

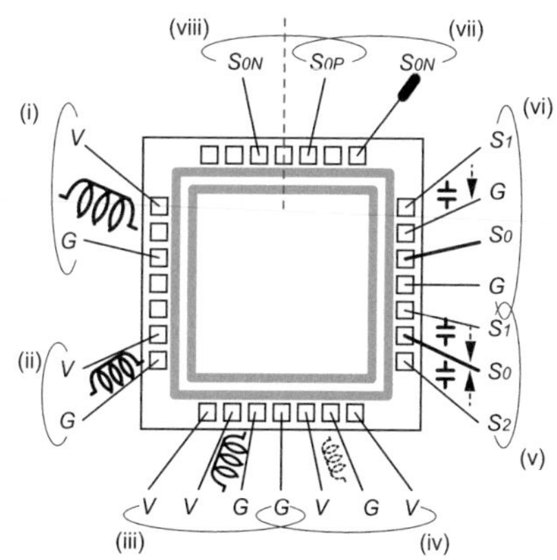

Fig. 5.16 Pin placement

5.4.3 Symmetry

The length of the bonding wires and lead frames differ from pad to pad, and thus not only do the signal transmission times from the package pin to the pad differ but also the voltage and current waveforms due to impedance mismatch. Therefore, for waveforms that change simultaneously, such as those of complementary signals, pins with symmetry in the package shape must be used, not as in (vii) but as in (viii).

Also, because the resistance and inductance of a bonding wire increase with the length, pins toward the middle of the chip with shorter bonding wires should be used for supply lines and important signals.

5.4.4 Assembly and Measurement

Pin placement also affects PCB board design. Things such as the routing of supply, ground, and signal lines on the board as well as the ease of assembly of external decoupling capacitances must also be considered. In addition, IO buffers should be selected with the characteristics of the measurement equipment and cables in mind, and the pin placement should be determined while considering the manageability of measurement.

Chapter 6
Noise

So you've finally made your LSI, but when you measure it, it's not showing any response... Consulting your senior is no good because he simply brushes it off as "noise is your problem." Here, "noise" does not mean "constantly random fluctuations such as thermal noise," but rather the word is used in a much broader sense, to mean "things that were unexpected during design."

6.1 Types and Causes of Malfunction

6.1.1 No Response

6.1.1.1 Measurement-Related Mistakes

The chip has arrived. The measurement setup is assembled, the power supply and input signals are connected, and waveforms are ready to be observed. However, no matter whether the supply voltage is raised, the CLK frequency is lowered, or any form of input is given, there is absolutely no response shown. It is even possible that when the power supply is turned on, a large current has flown and the chip has melted. Having said that, I have yet to experience observing the desired waveform on my very first try. The power supply connections should be verified, the ground connections should be verified, and the input signals should be directly observed on the oscilloscope to make sure they are correctly being fed in to the chip. Every possibility imaginable must be considered, and after trying even the wildest of ideas, it will still take at least one full day of headaches in the lab until a desired waveform is achieved. When the circuit is even slightly complex, trial and error can make a week easily go by. If the chips still show no response after this, then design mistakes are suspected.

T. Nakura, *Essential Knowledge for Transistor-Level LSI Circuit Design*,
DOI 10.1007/978-981-10-0424-7_6

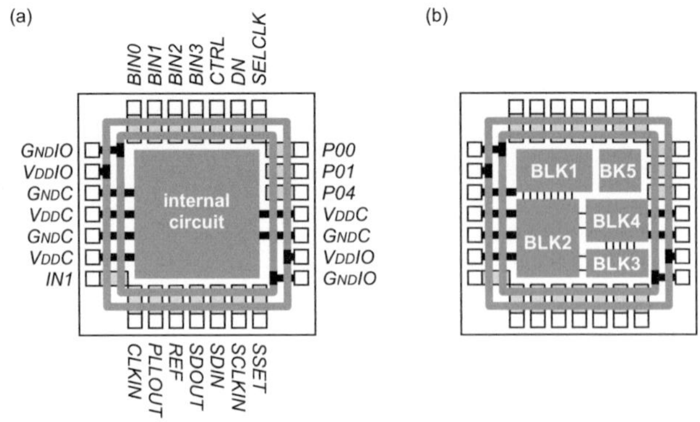

Fig. 6.1 Top level LVS/DRC of the chip

6.1.1.2 Design Mistakes

After reexamining the measurement setup and determining that the problem is not with the measurement, then there is potentially a mistake in the circuit design. This type of accident frequently occurs in university research labs where people have little design experience. The victims(?) can often be heard blabbering about how they never conducted a chip level LVS/DRC verification as in Fig. 6.1a, but rather only conducted block level LVS/DRC as in Fig. 6.1b, and "the rest of the connections are fine because they were visually verified."

Most occurrences of these accidents can be avoided by actually running good LVS/DRC. Always run LVS/DRC at the top level of the chip, as shown in Fig. 6.1a. Here, labels (e.g., IN1 or VDDC) should be drawn at the uppermost layer of the pads, and all other layers should not be referenced. Even if multiple people design separate blocks, and even if those blocks are not connected at all, an overall schematic of the entire chip must be made, and LVS must be run regardless. Also, when the layout is modified ever so slightly (e.g., if a width of a wire is changed from 0.11 to 0.10 μm), LVS/DRC must be run for verification, no matter how sure you are of the modification not being problematic.

6.1.2 Timing Errors

An error in the timing can occur due to the delay of a circuit changing from some effect. Changes in delay can occur as variation in logic delay or variation in the clock, and errors can occur as setup time violations or hold time violations.

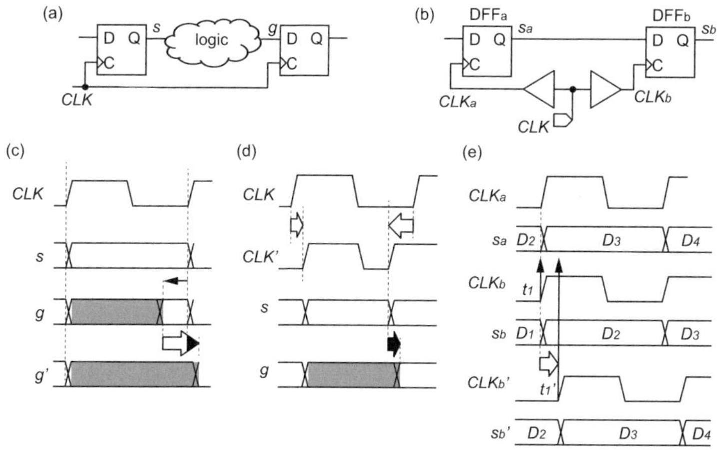

Fig. 6.2 Timing errors

6.1.2.1 Setup Violations

In a general synchronous circuit as shown in Fig. 6.2a, logic circuits comprised, for example, of NAND and NOR gates are placed between D flip-flop (DFF). In this structure, the delay from the DFF output s to the input of the DFF of the next stage g must be within one clock cycle. In other words, the logical value of g has settled before the rising edge of CLK, as shown in Fig. 6.2c. This time during which the logical value must settle is called the setup time. Here, when the delay of the combinational circuit increases due to the effects of noise and becomes as g' in Fig. 6.2c, the DFF is unable to latch the correct value, and the circuit will malfunction. This is called a setup violation. Also, as shown in Fig. 6.2d, a setup violation can occur when the CLK period changes due to the effects of noise, even if the combinational circuit delay does not change.

In a circuit with setup violations, the delay is too large relative to the clock period. Therefore, correct functionality can be obtained by increasing the supply voltage to reduce the logic delay or to lower the clock frequency.

6.1.2.2 Hold Violations

This can occur when the combinational circuit delay is too short. In a circuit as in Fig. 6.2b, CLK_a and CLK_b are input at the same time, and as indicated in Fig. 6.2e, DFF$_b$ will latch D_2, the output of DFF$_a$, at time t_1 and continue to output D_2 for one cycle. However, if the input clock timing of DFF$_b$ is delayed as in CLK_b' relative to the timing of CLK_a due to noise, DFF$_b$ will latch D_3 instead of D_2 at time t_1'. This type of error is called a hold error. Hold errors cannot be remedied by raising the supply voltage or reducing the clock frequency and is difficult to trace as a source of error, so it is important to be careful during design.

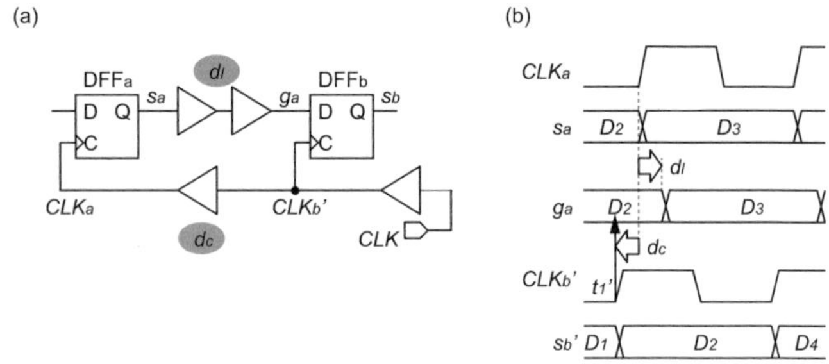

Fig. 6.3 Prevention of hold violations

As a measure against hold violations, a certain amount of delay should always be inserted between DFFs as shown in Fig. 6.3a. Also, if there is a DFF that may potentially cause hold violations, a delay to the clock signal can be inserted relative to the later DFF, if possible. With these techniques, the data arriving at DFF_b will come d_l later, and the latching instant of DFF_b will be d_c earlier, as shown in Fig. 6.3b, and the data D_2 can be latched safely.

6.1.3 Analog Errors

In analog circuits, it is often the case that the circuit has basic functionality but does not achieve the characteristics observed with simulations during design. For example, if an opamp is designed, there may be some amplification, but the gain might be only 20 dB instead of the expected 30 dB, the unity gain frequency was supposed to be 1 GHz but is actually only 900 MHz, and so on. The analysis of the cause of such errors is very difficult. The reason is first postulated, then a measurement strategy to confirm the hypothesis must be devised, and the measurement results are observed for further chasing of the reason of error. The process can be likened to invading a castle by burying the outer moat. Often, the root cause of the errors are the transistors operating in the linear region instead of the saturation region, due to various reasons such as process variation. For example, internally generated bias voltages might be inaccurate.

6.2 Types of Noise and Their Countermeasures

6.2.1 PVT Variations

PVT stands for process, voltage, and temperature.

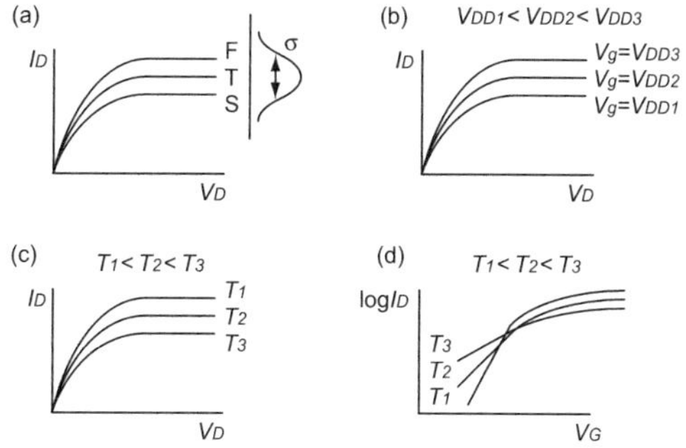

Fig. 6.4 Changes in transistor characteristics due to PVT variations

6.2.1.1 Process Variations

The temperature and humidity inside the foundry are maintained to be constant, but the manufactured transistor characteristics vary from time to time on the whim of a manufacturing device, and the characteristics between the SPICE parameters used for design and simulation and the actual fabricated chip differ. Usually, three types of SPICE parameters are provided, as indicated in Fig. 6.4a: Fast (F), Typical (T), and Slow (S). In regular logic circuits, a larger drain current makes the maximum speed of operation faster, thus the version with larger drain current is often called F, and the smaller version is called S. Also, the distribution of drain current is usually a normal distribution, but the SPICE parameters given in F and S can be at 1σ or at 3σ. Note that a 1σ variation means that 16 % of all chips are above F and 16 % are below S in variation. In Fig. 6.4a, the case for 1σ is shown. Also, sometimes five types of SPICE parameters (TT, FF, FS, SF, and SS) are provided. These are the cases when the PMOS and NMOS variation processes considered are common or separate. For example, if the gate oxide layer generated is thin (F), this is a common variation to PMOS and NMOS, because it is never the case that PMOS gates are thin (F) and NMOS gates are thick (S) simultaneously. In general, the F in FF is a stronger F than the F in FS. Relative to T, the drain currents in F and S often vary by about ±10 %.

6.2.1.2 Supply Variations

It is desirable to have a constant supply voltage for LSI, but in mobile devices the battery charge level will affect the voltage, and also temporary fluctuations in the voltage can also occur due to noise. As indicated in Fig. 6.4b, a larger current obviously flows when the voltage is higher, and the speed of operation becomes faster. In general, a stable operation is required even if the voltage varies from the standard value by ±10 %.

6.2.1.3 Temperature Variations

Transistor characteristics can change due to the temperature of the LSI environment as well. As shown in Fig. 6.4c, a larger saturation current flows when the temperature is lower. On the other hand, the leakage current increases when the temperature is higher. In other words, both the saturation current and leakage current are better when the temperature is lower. Functional operation is desired sometimes in the range of 0∼100 °C, and other times in the range of −25∼125 °C.

6.2.1.4 Corner Conditions

When the three types of variation P, V, and T are considered, there exist $2^3 = 8$ corner case conditions. In regular design, the typical conditions (TT/V_{DD}/27 °C) are initially used, after which simulations are used to confirm operation in the Fast (FF/1.1V_{DD}/−25 °C) and Slow (SS/0.9V_{DD}/125 °C) corners where transistors have the best and worst characteristics, respectively. The circuit is then adjusted to make sure that the desired operation is realized even under those corner conditions.

With digital circuits, the timing is designed so that setup violation errors do not occur in the Slow corner (SS/0.9V_{DD}/125 °C) and hold violation errors do not occur in the Fast corner (FF/1.1V_{DD}/−25 °C).

With analog circuits, the balance between PMOS and NMOS also becomes important, not just the Fast and Slow corners. For example, if an amplification of 100 times is desired for an input of $0.4 + 0.001 \sin \omega t$ under a 1.2 V supply voltage, a circuit such as the one drawn in Fig. 6.5a may be designed, and the PMOS transistor sizes might be adjusted so that the input-output characteristics of the first and second inverter stages become as (i_1) and (ii_1) as shown in Fig. 6.5b. While a clean amplification will be observed in simulation with typical conditions, unfortunately this circuit will never work when an actual chip is fabricated. When the NMOS is fabricated as a slightly Fast device during the fabrication process, the input-output characteristics of the circuit will become (i_2) and (ii_2) in Fig. 6.5b, and

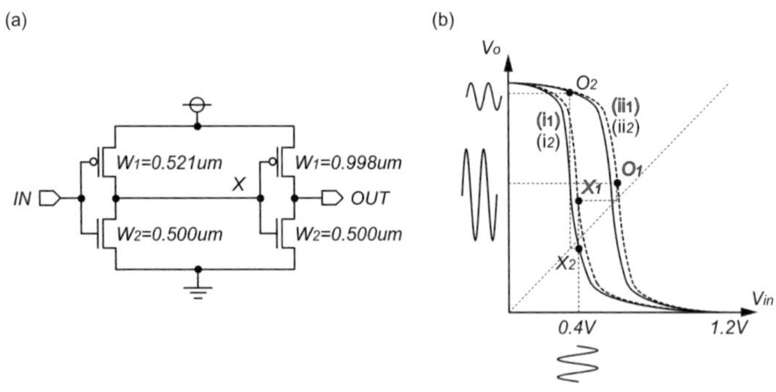

Fig. 6.5 Bias point change due to PVT variations

the output will operate at the bias point O_2; thus as a result, the signal will barely be amplified.

In the design of analog circuits, a circuit design where the bias point does not change by much even under PVT variations is necessary. For the validation of this, not only is confirmation necessary in the Fast and Slow corners, but all PVT corners such as the FS and SF corners must be simulated as well. However, because the typical conditions plus various corner simulations are laborious and time consuming, verification is often conducted with the five common corner cases.

6.2.2 *Supply Noise*

A large current runs through the supply, and noise is much more easily generated. The supply voltage fluctuates depending on the changes in the circuit current consumption, mostly due to noise sources such as the IR drop of the parasitic resistances and di/dt noise of the parasitic inductances in the supply lines. A schematic of the parasitic elements in supply lines is shown in Fig. 6.6a. Our concern is not with the fluctuation of the voltage relative to the ground outside of the chip, but rather with the voltage difference between the internal supply of the chip and the ground. Thus, Fig. 6.6a, b are equivalent. Here, how would the supply waveform appear internally to the chip? The voltage drop when a current I flows is $\Delta V = IR$ if the supply impedance is purely resistive and $\Delta V = LdI/dt$ if it is purely inductive, as shown in Fig. 6.6c, d. High-frequency components will be supplied by capacitors through insertion of decoupling capacitances C within the chip, and therefore the solid lines in Fig. 6.6c, d will turn into the supply voltage waveforms indicated by the dotted lines. In Fig. 6.6c, the average voltage drop does not change, but the absolute worst value is improved, thus avoiding chip malfunctions. In Fig. 6.6d, the

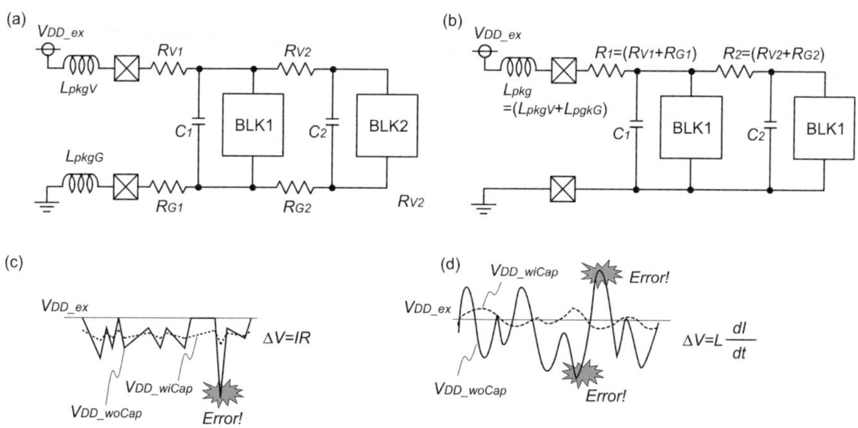

Fig. 6.6 Parasitic impedances and noise waveforms of supply lines

average value is V_{DD_ex} to begin with, but the fluctuations are more gradual and the worst case value is improved, thus avoiding chip malfunctions.

Effective methods to suppress supply noise are to reduce the parasitic resistance, to reduce the parasitic inductance, and to increase the decoupling capacitance. These methods have been explained previously but are summarized again below.

6.2.2.1 Parasitic Resistance

Parasitic resistances and contact resistances exist in packaging and bonding wires. As shown in Fig. 6.7a, the resistance is lowered by increasing the number of pins used for the supply ($R//R \rightarrow R/2$). Also, at locations within the chip that are further away from the supply pin, the resistance becomes larger, and voltage drops are more severe as shown in Fig. 6.7b. Supply pins should be spread out to make the longest distance to any internal point in a chip shorter.

Parasitic resistances also exist in the supply lines internal to the chip. As shown in Fig. 6.7c, supply lines should be as thick as possible in layout and should have a mesh structure. If possible, multiple layers should be placed as shown in Fig. 6.7d ($R = \rho l/wd$). When transitioning from layer to layer, as many vias as possible should be placed to lower the via resistance as shown in Fig. 6.7e, f ($R//R \rightarrow R/2$).

The root of supply pins carries the most current, and if there is a voltage drop at that point, then voltages at all the supply lines beyond that point will drop. It is

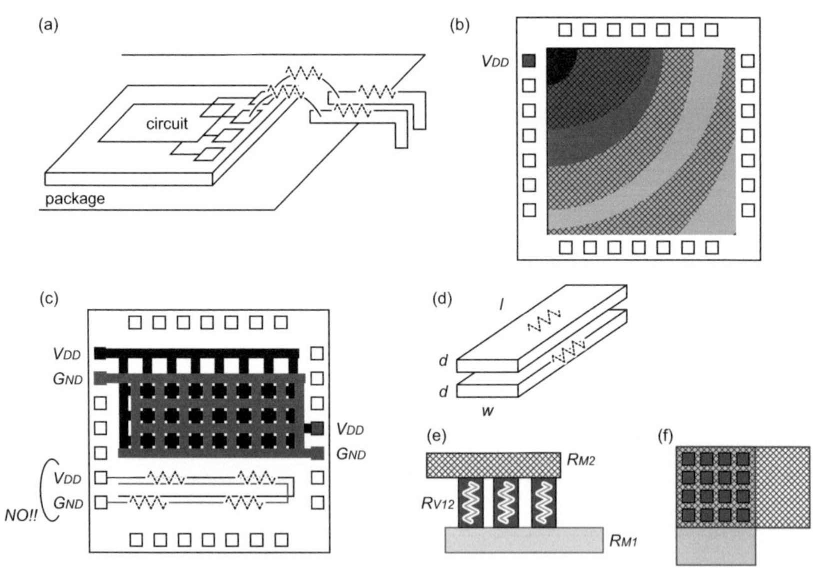

Fig. 6.7 Lowering the parasitic resistance of supply lines. (**a**) Multiple supply pins, (**b**) supply voltage drop map, (**c**) thicker supply lines, (**d**) multiple supply line layers, (**e**) (**f**) multiple vias

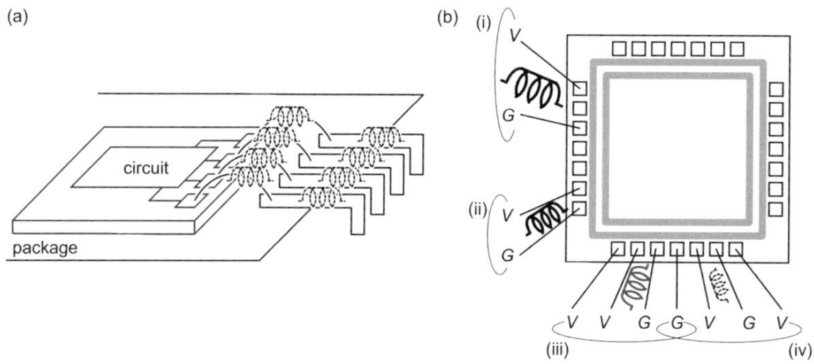

Fig. 6.8 Parasitic inductance of the supply lines

important to do the layout with special care so as to not cause voltage drops at these root points.

6.2.2.2 Parasitic Inductance

As shown in Fig. 6.8a, parasitic inductance exists in the package and bonding wires. The inductance is, as a rule of thumb, about 1 nH per 1 mm. The inductance due to the internal supply wiring of the chip can be ignored in most cases.

To make the inductance value smaller, the distance between supply and ground pins should be made smaller as shown in Fig. 6.8b (i) and (ii) ($\Phi = \int \boldsymbol{B} \cdot \boldsymbol{n} dS$, $L = \Phi/I \rightarrow L$ small when S small). Also, the inductance can be reduced by increasing the number of pins used for the supply ($L//L \rightarrow L/2$). In this case, the placement should not be VVGG as in (iii), but rather VGVG as in (iv) to make the inductance smaller.

6.2.2.3 Decoupling Capacitance

It is best to place as much capacitance between supply and ground as possible. As shown in Fig. 6.9a, b, PMOS and NMOS devices are turned on, and a capacitance is created between the gate and channel containing the gate oxide layer of the transistor. To make the capacitance per unit chip area larger, the source and drain regions can be reduced, and a larger gate can be used. However, there is channel resistance as shown in Fig. 6.9b, so this method reduces the capacitive characteristics at high frequencies. An appropriate gate size is determined by considering the relationship between the frequency of the noise to suppress and the time constant τ of the gate capacitance and channel resistance. Also, a gate leakage current I_{gleak} flows through the gate oxide layer in recent processes, and effects such as the reduction of battery life due to this leakage current must also be considered.

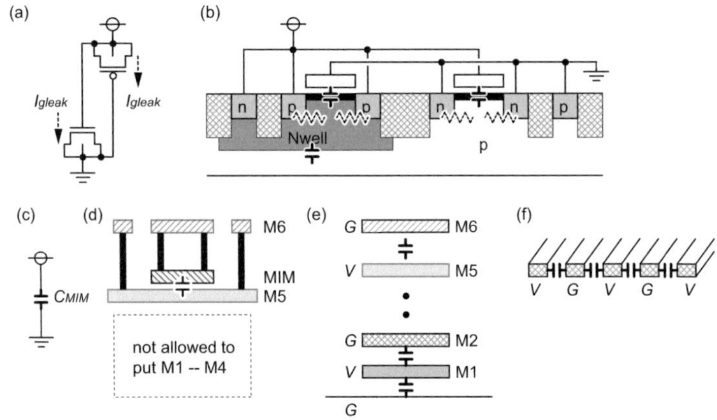

Fig. 6.9 Capacitance of supply lines. (**a**) (**b**) MOS gate capacitance, (**c**) (**d**) MIM capacitor, (**e**) (**f**) interconnect capacitance

The effects of leakage current diminish when a MIM capacitor as shown in Fig. 6.9c, d is used, but because the inter-MIM-layer insulator is thick compared to the gate oxide layer, a larger capacitance is more difficult to create ($C = \epsilon S/d$). Additionally, it is often prescribed that wires and transistors cannot be placed under MIM capacitors. The unit capacitance per area is thus less than 1/10 of that of a MOS gate capacitor.

Wiring capacitance can be added by overlaying the supply and ground interconnections as shown in Fig. 6.9e or by placing them next to each other in the same layer as shown in Fig. 6.9f.

These capacitances are combined in the remainder space of the chip layout to place as large a decoupling capacitance as possible.

6.2.3 Substrate Noise

In standard LSI, as shown in Fig. 6.10a, a large portion of the chip is taken by digital circuitry, while some parts are analog. Here, noise is generated by the switching of the digital circuitry, and this noise can propagate to the analog circuitry, causing the analog circuitry to malfunction. To suppress the transmission of noise, the digital circuit supply and the analog circuit supply are separated, as shown in Fig. 6.10d. However, noise can propagate through this following path: digital circuit G_{ND} – P substrate contact – Pwell – P substrate – Pwell – P substrate contact – analog circuit G_{ND}. This is called substrate noise. Substrate noise is quite a nuisance, and we often suffer due to this noise.

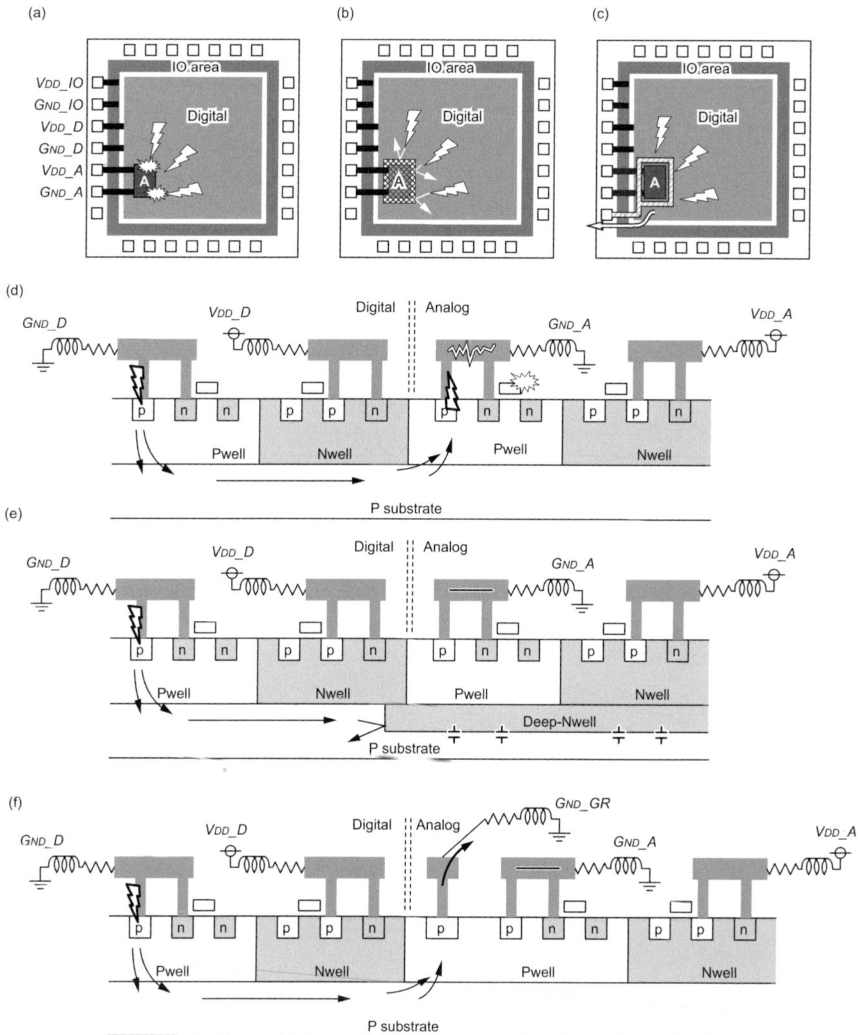

Fig. 6.10 The propagation and suppression of substrate noise

6.2.3.1 Deep N-Well

To split the path in the substrate through the P domain, a deep N-well, which is deeper than a regular N-well, can be used to surround the entire analog circuitry to prevent the transmission of substrate noise, as shown in Fig. 6.10b, e.

However, a PN junction capacitance exists between the deep N-well and the P substrate, and therefore, high-frequency substrate noise can still propagate through this junction capacitance.

6.2.3.2 Guard Ring

By surrounding the analog circuitry with P substrate contact and connecting this with outside ground as shown in Fig. 6.10c, f, substrate noise can be released outside before ever reaching the analog circuitry. This is called the guard ring.

However, noise will not escape to the outside if the ground line impedance for the guard ring is high, and noise that is not captured by the guard ring will still reach the analog circuitry.

6.2.4 Cross-Talk Noise

Parasitic capacitance exists between neighboring wires. As indicated in Fig. 6.11a, when the voltage on a node with a parasitic capacitance is changed, the voltage level of the neighboring wire is affected through the capacitance. This is called cross-talk noise, and the generator of noise is called the aggressor, whereas the receiver of noise is called the victim. In the wiring for analog circuitry whose operation depends on minute voltage changes, a design with special care for preventing cross-talk noise is necessary.

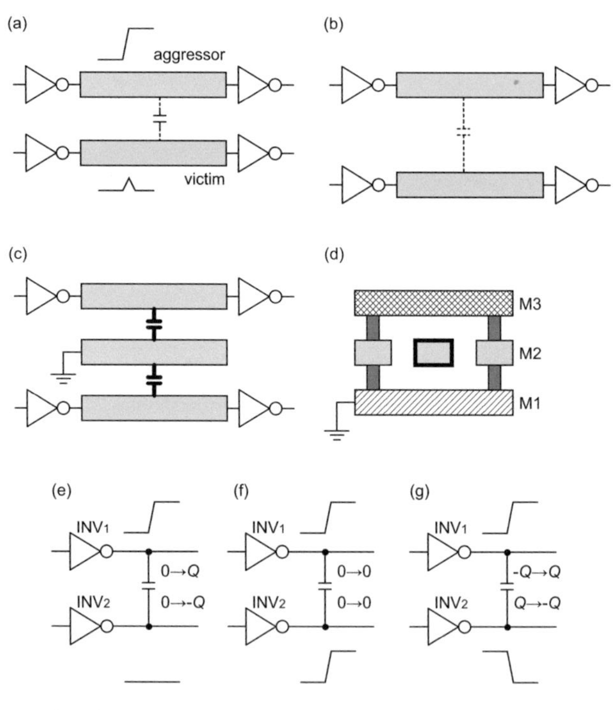

Fig. 6.11 Cross-talk noise

To prevent cross talk, the parasitic capacitance can be reduced by increasing the interconnect separation ($C = \epsilon S/d$) as shown in Fig. 6.11b, or a line with a fixed voltage such as ground can be inserted between wires as a shield as shown in Fig. 6.11c. For especially sensitive wires in analog circuitry, all four sides can be surrounded as shown in Fig. 6.11d to protect it from cross-talk noise.

Also, cross talk not only affects the voltage but also changes the propagation delay, which means it needs to be considered in digital circuit design as well. When the charge supplied from INV_1 to the parasitic capacitance while the voltage at the target does not change is Q as indicated in Fig. 6.11e, the charge supplied from INV_1 is zero when the target shows the same voltage change as shown in Fig. 6.11f and $2Q$ when the target has the opposite voltage change as shown in Fig. 6.11g. Thus, care is necessary because the delay of INV_1 varies as (f) < (e) < (g).

6.2.5 EMC

A magnetic field is generated in the surrounding area of a current running through a wire ($\nabla \times \boldsymbol{B} = \mu_0 \boldsymbol{i} + \mu_0 \epsilon_0 \frac{\partial \boldsymbol{E}}{\partial t}$). In the case of alternating current, the generated magnetic field also changes with time and correspondingly the electric field as well ($\nabla \times \boldsymbol{E} = -\frac{\partial \boldsymbol{B}}{\partial t}$). This change in electric field causes a change in the magnetic field. $\boldsymbol{E}$ and $\boldsymbol{B}$ induce each other and propagate through space. Thus, as shown in Fig. 6.12, an electromagnetic wave is radiated into space when an alternating current flows, and another LSI which receives this electromagnetic wave can malfunction. This is called electromagnetic interference (EMI), and the sensitivity of the receiver of EMI is called electromagnetic susceptibility (EMS). EMI and EMS together are called electromagnetic compatibility (EMC). The reason why people say "do not use cell phones inside of an airplane" or "do not use cell phones in a hospital" is that the strong radio waves purposefully emitted from cell phones turn into EMI for

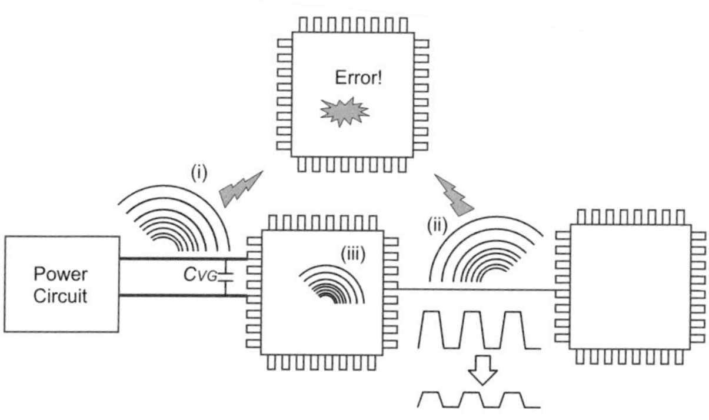

Fig. 6.12 EMC

LSI components in other devices, and this may cause malfunctions in LSI used in devices for airplanes or by hospitals and patients if the EMS was insufficient.

Some sources of EMI are (i) emission from supply lines, (ii) emission from inter-chip signal lines, and (iii) emission from inside a single chip. The largest source is (i), from supply lines. Because a large current flows from the power supply, the generated EMI is also large. By inserting a capacitance C_{VG} close to the supply pins of the chip, the AC components are supplied from C_{VG}, and the current coming from the power supply becomes a constant DC current, thereby suppressing the generation of EMI. EMI from signal lines (ii) can be suppressed by reducing the signal swing. In modern LSI sizes and operating frequencies, EMI from within a chip (iii) is not a significant problem.

As a general point of precaution, it is desirable not to create any wires that may resonate with the operating frequency f_{CLK} or the resonating frequency determined by parasitic inductances and decoupling capacitances ($f_{\mathrm{res}} = 1/2\pi\sqrt{LC}$). For example, the wavelength of 1 GHz in vacuum is 30 cm, so a wire of length 30 cm/4 = 7.5 cm should not be used or connected to an LSI that operates at 1 GHz. However, the dielectric constant of the board and the material inside LSI are not 1, so make sure to adjust accordingly.

Chapter 7
Problems Due To the Progress of Miniaturization

When you show your completed circuit diagram to your professor, he gives you a hard time: "what about the effects of variability?" and "have you considered NBTI?" What in the world are those?

7.1 Variation

7.1.1 About Variation

In Sect. 6.2.1, we emphasized the need for design with the effects of PVT variation on the final circuit characteristics taken into consideration. However, in recent processes, the changes in characteristics due to process (P) variation have become especially grave, and design is becoming much more difficult and painstaking. The transistor structure after the progress of device miniaturization is depicted in Fig. 7.1a, b. Even for the same gate length variation (ΔL), as miniaturization progresses, the fractional change ($\Delta L/L$) becomes larger. Even if the same defects occur in the gate oxide layer, their effects become larger. Also, the nonuniformity of the ions that determine P and N regions will have greater effects. As can be seen, the fabrication of uniform transistors becomes increasingly difficult as transistors become smaller, and individual transistor characteristics have come to vary wildly. This is called "variation."

7.1.2 Types and Causes of Variation

Variation has its types. As shown in Fig. 7.2, some types of variation are (b) inter-wafer variation, (c) intra-wafer or inter-chip variation, (d) intra-chip variation that

T. Nakura, *Essential Knowledge for Transistor-Level LSI Circuit Design*,
DOI 10.1007/978-981-10-0424-7_7

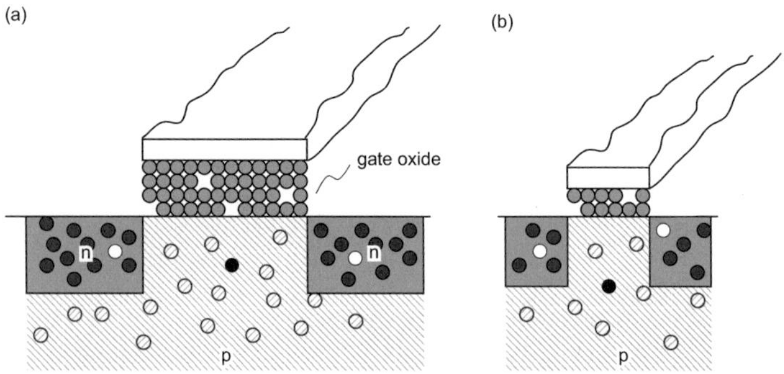

Fig. 7.1 The occurrence of variation

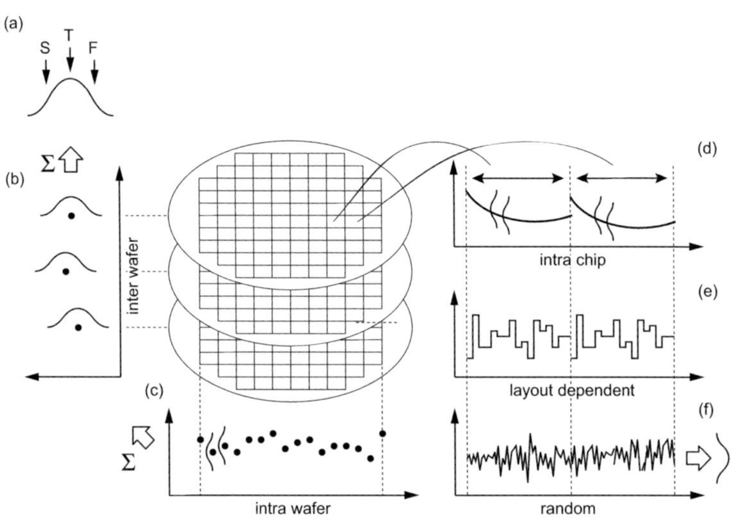

Fig. 7.2 Variation

depends on the location within a single chip, (e) layout-dependent variation, and (f) random variation of each individual transistor. The combination of each of these variations becomes (a) the actual transistor variation.

The reasons for these variations are listed below:

(a) Transistor variation: the combination of (b)–(f) below
(b) Inter-wafer variation: disturbance in the stability of the manufacturing equipment
(c) Inter-chip variation: disturbance in the uniformity of the manufacturing equipment
(d) Intra-chip variation: nonuniformity of lithography, dependence of patterning on the annealing temperature

(e) Layout-dependent variation: proximity effects of lithography (the limits of OPC)
(f) Random variation: the nonuniformity of dopant concentrations, or random dopant fluctuation (RDF)

Here, the inter-wafer (b) and inter-chip (c) variations are called "global variations," in the sense that these are uniform variations where all transistors on the same chip have the same characteristics. The manufacturing variation (TT, FF, SF, etc.) mentioned in Sect. 6.2.1 is global variation. On the other hand, intra-chip (d), layout-dependent (e), and random (f) variations are called "local variations" in the sense that even transistors of the same size within the same chip can have varying characteristics. When local variations exist, SPICE simulations cannot be conducted during design, which causes difficulty. The random variation (f) is especially difficult to deal with because transistors with the same layout that are side by side can have different characteristics.

As a guideline for the magnitude of random variation, Pelgrom's relationship

$$\sigma_{Vt} = \frac{A_{Vt}}{\sqrt{LW}} \tag{7.1}$$

is well known. This indicates that the transistor threshold voltage variation is inversely proportional to the square root of the gate area. This matches with the intuition that the average value of the dopant concentration approaches a constant as the gate area increases.

7.1.3 *The Effects of Variation*

7.1.3.1 SRAM

SRAM contains more analog-like elements than logic circuits and thus is more susceptible to variation effects. Also, the symmetric structure is laid out in a large quantity, allowing the direct observation of the effects of variation and leading to the ease of measurements. Due to such characteristics, SRAM is most often used for experiments regarding variation. Here, the operating principles of SRAM are explained enough so that literature regarding variation can be understood.

In the operation of SRAM, errors occur most frequently not during write-in but during readout. The SRAM structure and the timing chart during a readout are as shown in Fig. 7.3a, b. An SRAM cell consists of cross-coupled inverters, the word line *wl* that controls access to the cell, and the bit lines *b* and *bb* that are the data inputs and outputs. During a read operation, initially the bit lines *b* and *bb* are precharged to ONE. Here, there is a buildup of charge on the relatively large wiring parasitic capacitance C_b on the bit lines. Assuming that the nodes *q* and *qb* are holding the values ONE and ZERO, respectively, when the readout signal on the word line *wl* is turned into a ONE, M_4 will try to pull down the potential

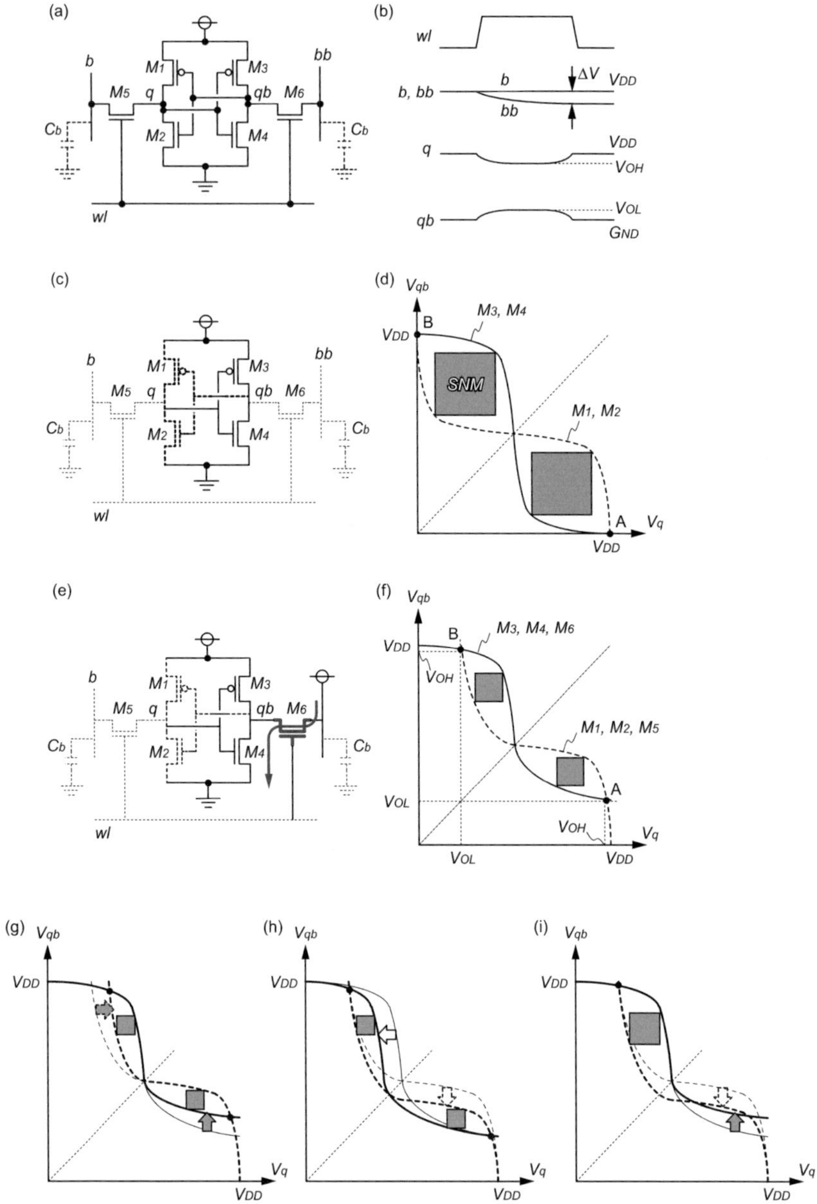
(a)
(b)
(c)
(d)
(e)
(f)
(g)
(h)
(i)
b
bb
M1
M2
M3
M4
M5
M6
q
qb
Cb
wl
b, bb
ΔV
VDD
VOH
VOL
GND
Vqb
Vq
M3, M4
M1, M2
SNM
A
B
M3, M4, M6
M1, M2, M5

Fig. 7.3 SRAM

on bb through M_6. Then, the potential difference between b and bb can be observed through a sense amp to figure out that q held the value ONE and qb the value ZERO. Here, the potential on qb is not kept at the G_{ND} potential, but rises to a voltage V_{OL} which is determined by the ratio of on resistances of M_4 and M_6. Therefore, the output q of the left inverter M_1, M_2 will drop from V_{DD} to V_{OH}. To prevent these, the gate widths W of M_2 and M_4 are designed to be 2∼4 times greater than those of M_5 and M_6. However, this increases the overall area and the number of bits per unit area decreases.

Let's look at the characteristics of the SRAM cell inverter. When the word line wl is ZERO and M_5 and M_6 are off, the voltages V_q and V_{qb} at the input and output (q, qb) of the inverter of M_3 and M_4 in Fig. 7.3c have the CMOS inverter relationship drawn as the solid line in Fig. 7.3d, and likewise the input-output characteristic of M_1 and M_2 is the dotted line in Fig. 7.3d.

However, when M_5 and M_6 turn on during SRAM readout, the relationship between V_q and V_{qb} created by M_3, M_4, and M_6 are the curves shown in Fig. 7.3f. Even if V_q rises and M_3 and M_4 turn off and on, respectively, there is a current flow from M_6 that causes the voltage of V_{qb} to not be G_{ND} but to rise to V_{OL} even if V_q is at V_{DD}, creating the curve shown as the solid line in Fig. 7.3f. Similarly, the curve for M_1, M_2, and M_5 is shown as the dotted line in Fig. 7.3f. As a result, when a ONE and ZERO are stored on nodes q and qb within the cell, the voltage on q and qb become V_{OH} and V_{OL} as can be seen from these curves.

As for these curves shown in Fig. 7.3d, f, the more separated these two curves are, the more stable a readout is possible. This separation is called the static noise margin (SNM). When the transistor characteristics vary, these curves also vary. For example, if M_2 and M_4 both become weak, the characteristics become as that in Fig. 7.3g, and if M_1 and M_3 become weak, the characteristics become as that in Fig. 7.3h. In these cases, SNM is still ensured. However, if M_4 and M_1 remain the same but M_2 and M_3 become weaker due to random variation, the characteristics become as that shown in Fig. 7.3i and the cell fails. For example, if a 1 Gb (one billion bits) SRAM is created, some of these will have the characteristics shown in Fig. 7.3i, which leads to malfunctions. From this, local variations and random variations especially in the case of SRAM, rather than global variations, can be understood to be a serious problem leading to malfunctions.

7.1.3.2 Replicas

Due to the effects of variation, characteristics can differ even between copies (replicas) of the exact same circuit operating in the exact same manner.

The longest path in a logic circuit is called the critical path, and if the delay of the critical path falls within one clock period, all logic delays must fall under one clock period. For example, when the supply voltage and CLK frequency are controllable as with dynamic voltage and frequency scaling (DVFS), a copy of the critical path within the main circuit with some margin is separately prepared as shown in Fig. 7.4a. From this, the CLK frequency can be raised to the limit by

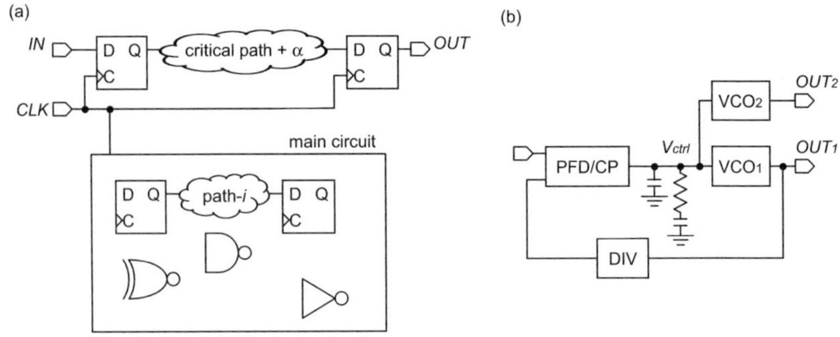

Fig. 7.4 Replicas

observing the operation of the replica circuit. This method is effective when the transistor characteristics within a chip are uniform, but if due to local variations the replica circuit happened to be manufactured at FF and the internal circuit path i was manufactured at SS, the delay of path i can exceed one clock period and cause errors.

In the case of analog circuits, an example is shown in Fig. 7.4b, where the frequency generated by VCO_1 is controlled by adjusting the control voltage V_{ctrl} through feedback to ensure the desired output frequency at OUT_1. Even in this case, if the entire circuit varies uniformly, the feedback will compensate for the variation and the desired oscillation frequency will be achieved, but due to random variation, the generated frequency of VCO_2 will differ from that of VCO_1 even if the exact same control voltage is applied.

7.1.3.3 Symmetric Analog Circuits

In an amplifier circuit as shown in Fig. 7.5, the current source I_0 is utilized so that the amplifier characteristics are not affected by global variations. By copying this current with current mirrors M_1, M_2, and M_7, the current I_0 also runs through M_2 and M_7, and the same desired characteristics are supposed to be achieved in AMP_1 and AMP_2. However, due to local variations, the characteristics of M_1, M_2, and M_7 differ, which causes the currents running through M_2 and M_7 to deviate from I_0, leading to varying amplifier characteristics between AMP_1 and AMP_2. Also, when M_3 and M_4, as well as M_5 and M_6, are the same size, as long as IN_1 and INB_1 are the same voltage, the output OUT_1 will be neither ONE nor ZERO, but rather some intermediate voltage will be output. However, if due to random variation an M_6 that has better performance than M_5 is manufactured, the output OUT will be at a value closer to V_{DD} even when IN_1 and INB_1 are at the same voltage, leading to an offset.

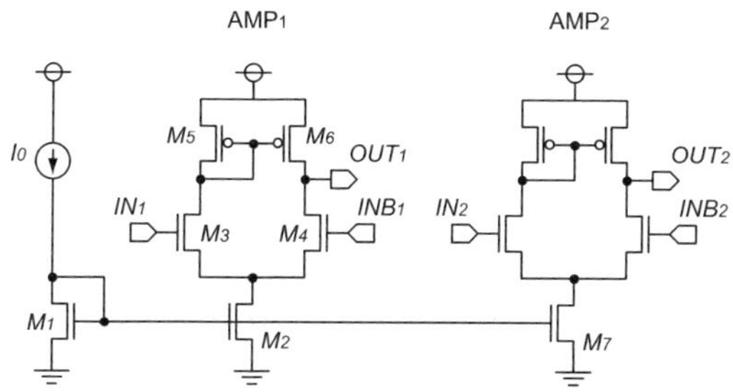

Fig. 7.5 Symmetry

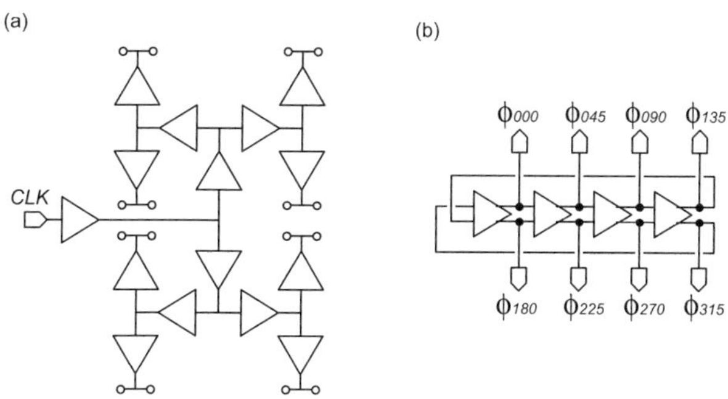

Fig. 7.6 (**a**) Clock skew, (**b**) phase offset

7.1.3.4 Clock Skew and Multiphase Outputs

To distribute a clock to a large-scale circuit, it is common to utilize an H-tree structure as shown in Fig. 7.6a. By making the number of buffer stages from the *CLK* pin root to the end leaves the same, the CLK rising-edge timing at the flip-flops connected to the end leaves is lined up. However, when the delay of each buffer varies due to local variations, this brings about variations in the input timing of flip-flops. This is called clock skew. As a result of this, the effective clock frequency is shortened and malfunctions occur due to setup and hold time violations.

Sometimes, signals that are offset by 45° relative to another signal oscillating at some frequency f_0 is required. By using multiple outputs of a buffer circuit as shown in Fig. 7.6b that outputs complementary signals, an 8-phase signal $\phi_{000} \sim \phi_{315}$ can be created. However, when the buffer delays vary due to local variations, each phase will deviate from perfect 45° steps (e.g., 0, 41, 97, 142, …).

7.1.4 Monte Carlo Simulations

As for global variations, model parameters are provided as libraries such as TT, FF, and FS, within the SPICE parameter file, and these can be utilized to run

SPICE simulations. What should we do for local variations? For example, to observe threshold variations, we could slightly change the size of each transistor in the netlist. However, if we have `.SUBCKT VCO`, then the characteristics of `XVCO1` and `XVCO2` cannot be made different. It also seems that LVS would become a pain.

To observe local variations, we conduct Monte Carlo simulations. This is when the SPICE simulator randomly varies specified parameters within the netlist when running simulations. The HSPICE netlist to run Monte Carlo simulations to simulate the characteristics of $V_q - V_{qb}$ during an SRAM cell readout as shown in Fig. 7.3a is as shown below.

```
*--------------------- Circuit Definition ---------------------
.OPTION POST=2 POST_VERSION=2001
.OPTION PROBE
.PROBE DC V(*) I(V*)

.TEMP = 27
.param mvdd = 1.2

VV  V  0 DC mvdd
VG  G  0 DC 0
VB  B  0 DC mvdd
VBB BB 0 DC mvdd
VWL WL 0 DC mvdd
*VWL WL 0 DC 0

m1 q qb v v P L=65e-9 W=800e-9
m2 q qb g g N L=65e-9 W=400e-9
m3 qb q v v P L=65e-9 W=800e-9
m4 qb q g g N L=65e-9 W=400e-9
m5 b wl q g N L=65e-9 W=200e-9
m6 qb wl bb g N L=65e-9 W=200e-9

*--------------------- SPICE parameters ----------------------
.param
+ sigma3_dvthn_g  = 0.2
+ sigma3_dvthn_l  = 0.05
+ sigma3_dvthp_g  = 0.2
+ sigma3_dvthp_l  = 0.05

.LIB NT
.param
+ dvthn_g = 0
+ dvthn_l = 0
.ENDL

.LIB NF
.param
+ dvthn_g = '-sigma3_dvthn_g'
+ dvthn_l = '-sigma3_dvthn_l'
.ENDL

.LIB NS
```

```
.param
+ dvthn_g = sigma3_dvthn_g
+ dvthn_l = sigma3_dvthn_l
.ENDL

.LIB PT
.param
+ dvthp_g = 0
+ dvthp_l = 0
.ENDL

.LIB PF
.param
+ dvthp_g = '-sigma3_dvthp_g'
+ dvthp_l = '-sigma3_dvthp_l'
.ENDL

.LIB PS
.param
+ dvthp_g = sigma3_dvthp_g
+ dvthp_l = sigma3_dvthp_l
.ENDL

.MODEL N NMOS
+ LEVEL = 1
+ VTO = '0.3 + dvthn_g + dvthn_l'

.MODEL P PMOS
+ LEVEL = 1
+ VTO = '-0.3 - dvthp_g - dvthp_l'

*--------------------- Simulation Control --------------------
VQ  Q  0 DC 0
*.DC VQ 0 mvdd 0.01
.DC VQ 0 mvdd 0.01 SWEEP MONTE=100 FIRSTRUN=1

.lib "sramcell.inp" NT
.lib "sramcell.inp" PT

*.param dvthn_g = AGAUSS(0, sigma3_dvthn_g, 3.0)
*.param dvthp_g = AGAUSS(0, sigma3_dvthp_g, 3.0)

.OPTION MODMONTE=1
.param dvthn_l = AGAUSS(0, sigma3_dvthn_l, 3.0)
.param dvthp_l = AGAUSS(0, sigma3_dvthp_l, 3.0)

.END
*-------------------------------------------------------------
```

In the beginning portion, terminal voltages and circuits are defined. In the section `SPICE parameters`, the variation parameters and their magnitudes are declared. Here, global variation variables `***_g` and local variation variables `***_l` are separately declared, the magnitude of variation is set to 3σ, the variation parameter values are set from libraries such as `NT`, `NF`, and the section `.MODEL` sets values for model parameters (here, the LEVEL1 model is used for simplicity).

The Monte Carlo simulation is managed in section `Simulation Control`. Here, a DC simulation is conducted, in which the voltage of q is swept from 0 to `mvdd` (set to 1.2 V in the beginning section) in 0.01 V steps. By adding `SWEEP MONTE=100 FIRSTRUN=1` to the line starting with `.DC`, the Monte Carlo simulation is run 100 times. Because the random number seed used in the Monte Carlo simulation is fixed each time, today's first to hundredth Monte Carlo simulation and tomorrow's first to hundredth Monte Carlo simulation will have the exact same results. If the line is changed to `FIRSTRUN = 101`, a set of random numbers from index 101–200 are generated, and a different (but statistically the same) result will be obtained. With `.OPTION MODMONTE=1`, the simulation becomes that for local variations where each transistor carries a different value. Without this line, the first to hundredth simulations will have different random numbers, but all transistors will have the same value and thus the simulation will be that for global variations.

With `.param dvthn_l = AGAUSS(0, sigma3_dvthn_l, 3.0)`, `dvthn_l` is defined to have a Gaussian distribution centered at zero and with 3.0σ of `sigma3_dvthn_l` (set to 0.02 earlier). `dvthn_l` is also defined in `.LIB`, but because HSPICE allows overwriting of parameters by redefining them, the variable can be set to have a Gaussian distribution by defining it after `.LIB`. In this example, global variations are fixed to be `NT`, `PT` conditions, and only local variations are simulated with Monte Carlo.

```
.param dvthn_g = AGAUSS(0, sigma3_dvthn_g, 3.0)
.param dvthp_g = AGAUSS(0, sigma3_dvthp_g, 3.0)

*.OPTION MODMONTE=1
*.param dvthn_l = AGAUSS(0, sigma3_dvthn_l, 3.0)
*.param dvthp_l = AGAUSS(0, sigma3_dvthp_l, 3.0)
```

With such a change, the Monte Carlo simulation becomes that with only global variation taken into consideration (intra-circuit parameters are uniform). It is (probably) not possible to make global variations uniform within a circuit and make local variations individual within a circuit. Also, it is probably more natural to set the local variation variable (`***_l`) inside `.LIB` to be zero.

7.2 Leakage Currents

It used to be stated often that the current consumption of LSI is the short circuit current during switching and the charge-discharge current of the capacitance of the next stage. However, in the devices of recent years, the leakage current has become large enough that it cannot be ignored. Because the leakage current directly impacts the standby time of mobile devices, techniques to reduce the leakage current are becoming necessary.

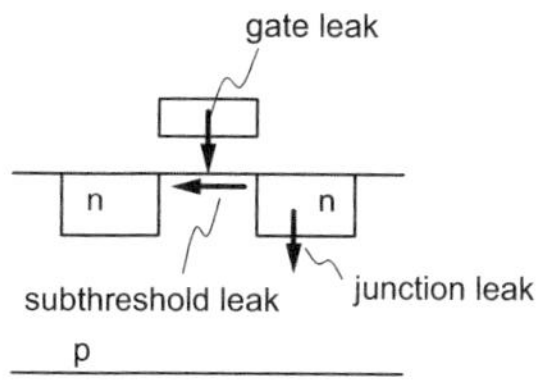

Fig. 7.7 Types of leakage current

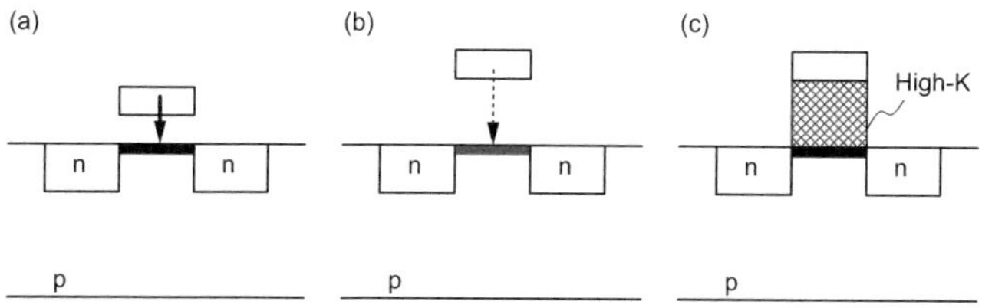

Fig. 7.8 Gate leakage

As shown in Fig. 7.7, the types of leakage current are subthreshold leakage, where a current runs between the source and drain even when the transistor is off; gate leakage, where current flows through the gate oxide layer; and junction leakage, where current runs in the reverse direction at the PN junction. In 45 nm technology, it is said as a rule of thumb that the ratio of gate leakage to subthreshold leakage to junction leakage = 5:4:1.

7.2.1 Gate Leakage and High-K

As miniaturization progresses and the gate oxide layer becomes only several nm thick, tunneling currents start to flow through the gate oxide, as shown in Fig. 7.8a. If the gate oxide is made thicker to suppress this tunneling current as shown in Fig. 7.8b, the drain current decreases and the device operation slows down. By using material with a high permittivity (high-K) for the gate oxide as shown in Fig. 7.8c, the gate oxide can be thickened to reduce the tunneling current, and simultaneously enough carriers are formed in the channel ($C = \epsilon S/d$) and a sufficient drain current is allowed to flow.

7.2.2 Subthreshold Leakage

The subthreshold current is $I_D = I_0(e^{qV_g/nkT} - 1)$, and the factor S, which is shown in Fig. 7.9b and is the inverse of the slope of $\log I_D$, is $S = nkT/q$ and does not depend on the process. In the ideal case of $n = 1$, this factor becomes 60 mV/decade, which

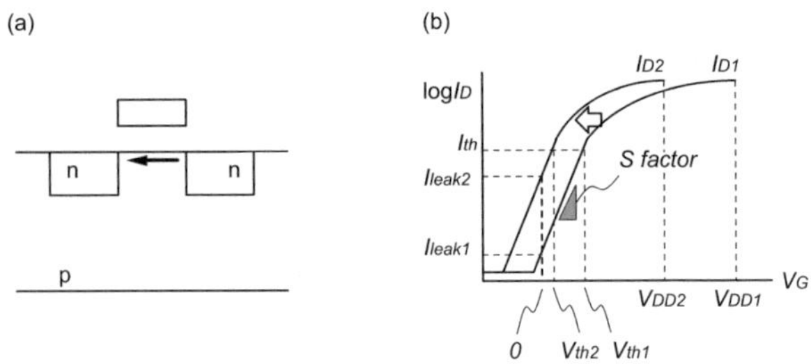

Fig. 7.9 Subthreshold leakage

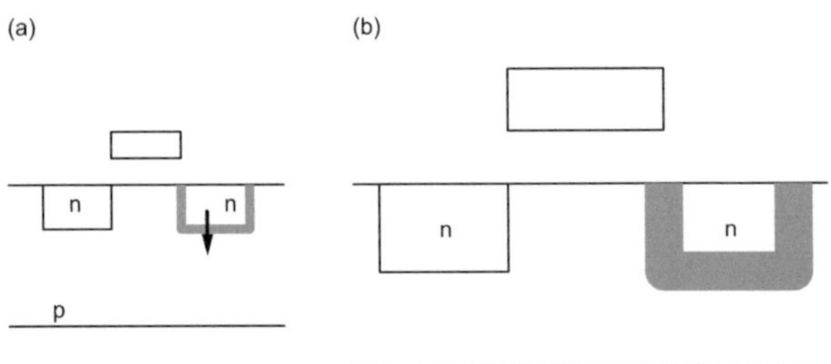

Fig. 7.10 PN junction leakage

means that if V_G decreases by 60 mV, the drain current becomes 1/10. In reality $n = 1$ does not occur and S becomes 70 or 80 mV/decade. This means that, for example, in the case of 70 mV/decade, if the threshold voltage decreases by 70 mV, the leakage current becomes 10 times greater.

As the process miniaturization advances and the supply voltage decreases, the threshold voltage must accordingly be set lower. Figure 7.9b indicates typical examples of transistor characteristics for when the supply voltage is V_{DD1} and in a finer technology where the supply voltage is V_{DD2}. Even if the supply voltage reduces the same drain current, I_D can be achieved, but because the subthreshold slope (S factor) that indicates the conditions below the transistor threshold does not change, the leakage current at $V_G = 0$ V is increased.

To reduce the leakage current, techniques such as sacrificing speed by using transistors with higher thresholds for circuit design and using multi-threshold circuits (MTCMOS) and power gating unused blocks are becoming necessary.

7.2.3 *Junction Leakage*

As transistor sizes become smaller, the dopant concentrations of the source and drain regions are increased in order to maintain their relative sizes. As a result, the depletion region of the reverse-biased PN junction becomes smaller as shown in Fig. 7.10, and a leakage current flows.

7.3 Degradation of Characteristics

Recent devices change in characteristic as they are used, and after a while, their characteristics can even degrade so much that they cause malfunctions. It is necessary to make the circuit structure such that the characteristics are less likely to degrade, or that even if they do, the overall circuit still functions.

7.3.1 Electromigration

Given the wires in Fig. 7.11a, b, which is more likely to break? With the progress of miniaturization, wires become thinner, and as a result, the current densities of the currents that run through these wires become higher. When these currents flow inside wires, the electrons collide with metal atoms as they proceed, and as shown in Fig. 7.11c, the electrons slowly displace the metal atoms, and the wire resistance starts to increase as a result of the uneven metal atom arrangement. This can even result in the breakage of the wire, and this is called electromigration. The larger the current density, the more likely electromigration is to occur. Copper is less prone to electromigration than aluminum, but that does not mean that electromigration does not occur at all. Also, this does not only occur in the wiring but also can occur in the vias that connect wires together.

Because electromigration occurs due to the displacement of metal atoms from the electrons moving in one direction, it is more likely to occur when a constant current is flowing in the same direction, such as the circuit which constantly outputs a ONE as shown in Fig. 7.11d. It is less likely to occur when the currents are time limited and bidirectional, as in Fig. 7.11e. Constant, unidirectional currents are common in analog circuits, and in these cases, the wiring should be made wide, as in Fig. 7.11f.

The maximum allowable current densities are defined in design manuals. For example, for a rule of 1 mA/μm, a current greater than 1 mA for every 1 μm of wire in width is not allowed. In this case, to flow a 2 mA of current, the wire width must be made greater than 2 μm. For vias, a rule of 0.1 mA means that a current greater

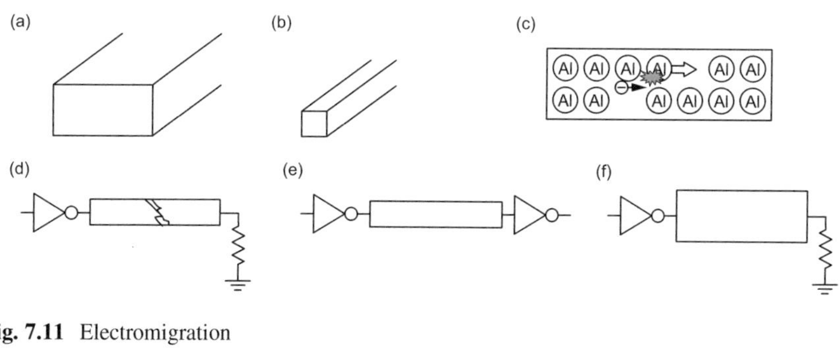

Fig. 7.11 Electromigration

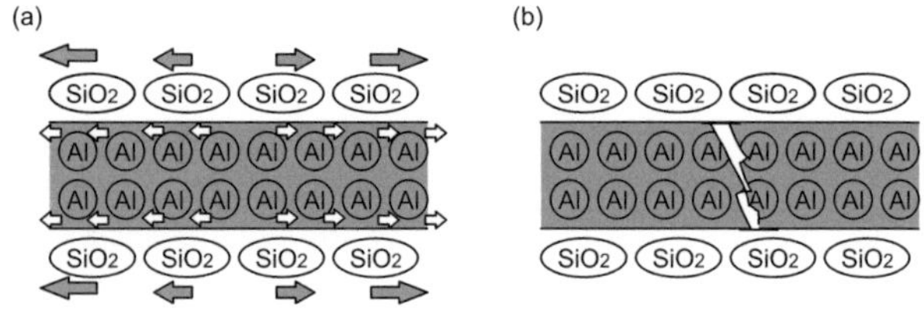

Fig. 7.12 Stress migration

than 0.1 mA per via is not allowed, so, for example, if 10 mA of current is desired, at least 100 vias must be placed in parallel. The distinction between DC and AC currents is often not made in design manuals.

7.3.2 Stress Migration

Matter expands when temperature rises. Because the thermal expansion coefficients differ due to the material, stress occurs at the interface between differing materials when the temperature changes. Within LSI, as shown in Fig. 7.12, the difference in expansion coefficients between the oxide layer and the metal wiring causes the wiring to be pulled, and the wire can increase in resistance or even break. This is called stress migration. The thinner the wire, the more likely this is to occur, and also copper wires are more prone than aluminum wires to this effect. Thus, this problem becomes more serious with the progress of miniaturization. However, as designers, we must simply design according to the wire thicknesses specified in the design rules, and mitigations to this problem are left to process engineers.

7.3.3 Soft Errors

Cosmic rays from space falling into Earth contain neutrons and α particles. In addition, the packaging material also contains trace amounts of uranium that emits α rays. When these α rays with high energy are injected into LSI, electron-hole pairs are generated as shown in Fig. 7.13a. The electrons and holes generated here will move according to the potential due to the transistor bias voltages. For example, when α rays are injected into DRAM, as shown in Fig. 7.13a, the generated electron-hole pair can potentially invert the charge stored on the DRAM capacitance. With the progress of miniaturization, the absolute amount of charge stored on the capacitor also decreases, which makes soft errors more likely to occur. Also, although soft errors previously only occurred in DRAM and not in logic circuitry, in recent years, SRAM and combinational circuits as shown in Fig. 7.13b are also prone to soft errors.

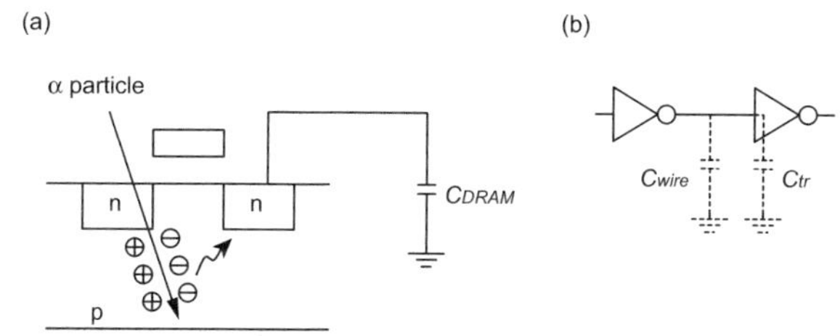

Fig. 7.13 Soft errors

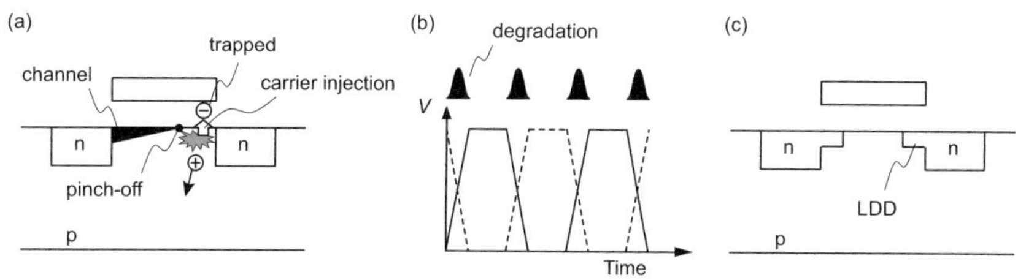

Fig. 7.14 Hot carrier injection

As measures against such errors, the chip (package) surface is coated with special material to prevent the injection of α rays, or defects are inserted on purpose in the silicon substrate so that the electron-hole pairs generated by α rays are absorbed. Besides such process techniques, there are circuit techniques to improve soft error resiliency, such as redundancy in flip-flops or enlargement of transistors in regions susceptible to soft errors.

7.3.4 Hot Carrier Injection

In a transistor operating in the saturation region, there exists a region with high electric field on the drain side of the pinch-off point at the edge of the channel. Within the channel, the electrons are accelerated by the electric field, but the electrons that have obtained energy in this high electric field region cause impact ionization, generating electron-hole pairs. Some of these electrons get trapped by the gate oxide layer and become fixed charge, as shown in Fig. 7.14a. This causes the NMOS threshold voltage to rise and correspondingly a degradation of the transistor characteristics. This is called hot carrier injection (HCI).

Hot carriers are generated when a current flows in the saturation region. Therefore, this effect is more likely to occur when the gate voltage is somewhat lower than V_{DD} and the drain voltage is at V_{DD}. In logic circuits, this occurs when a switching current flows. For example, let Fig. 7.14b be a CMOS inverter input-output characteristic. Because the characteristic degradation due to hot carriers occurs when the input and output change, HCI is less problematic at lower frequencies and more problematic at higher frequencies. This is becoming an even greater of a problem due to the combined effects of acceleration of operation frequency due to advanced LSI processes, the increase in threshold variation per electron due to the decrease in threshold voltages, and the increase in the electric field magnitudes due to the relative differences in scaling of the physical dimensions and the voltage levels.

In general, degradation of transistor characteristics occurs more often at higher temperatures, but in the case of hot carriers, the degradation is worse for lower temperatures. This is because as the temperature decreases, the heat vibrations of the lattice decrease and the probability of electrons colliding with the lattice also decreases, therefore increasing the probability of electrons becoming a hot carrier with high energy.

To prevent the degradation of characteristics due to hot carriers, a structure as shown in Fig. 7.14c with a lightly doped drain (LDD) is used. As for design techniques, bias states which would generate hot carriers should be avoided, and a circuit structure that would still operate even with degradation due to HCI is required.

7.3.5 NBTI

When a negative gate bias is applied to a PMOS device for an extended period of time, the threshold increases and the transistor characteristics degrade. This is called negative bias temperature instability (NBTI). The cause of NBTI is shown in Fig. 7.15. The gate oxide layer interface exists as Si-H, and when a negative bias is applied to the gate, a channel of holes is formed. The holes and Si-H cause an electrochemical reaction which releases hydrogen, generating traps within the oxide layer. The increase of this interface state as well as the traps in the oxide layer causes the threshold to rise. In recent years, the positive bias temperature instability (PBTI) of NMOS devices is being observed, not only the NBTI for PMOS.

Also, as a trait of NBTI, the device characteristics recover by applying a positive bias to the gate. As shown in Fig. 7.15b, the transistor characteristics return to nor-

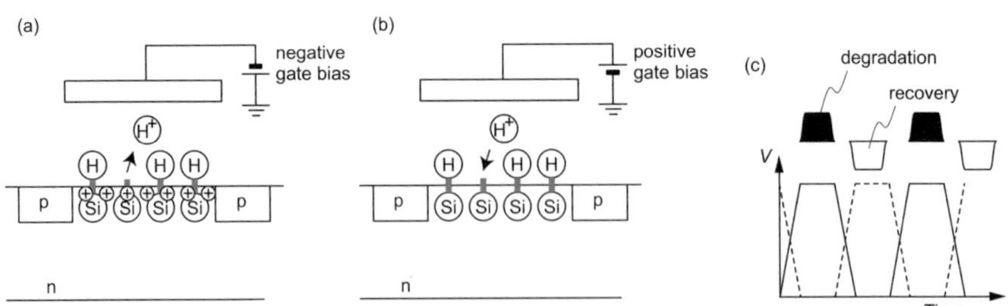

Fig. 7.15 NBTI

mal when the released hydrogen atoms return to the Si-H state. Therefore, for example, if Fig. 7.15c indicates CMOS inverter input-output characteristics, the NBTI lifetime is improved when alternating positive and negative biases are repeated because the device enters the degradation or recovery modes while the input and output stay HIGH or LOW, compared to applying a constant DC bias. As for circuit design techniques, along with designing circuitry that functions even with the degradation of transistor characteristics due to NBTI, there is potential to prevent such degradation due to NBTI by leveraging the recovery trait in the circuit structure.

7.3.6 *Random Telegraph Noise*

There exist states that trap charge in the interface to the gate oxide layer. As shown in Fig. 7.16a, b, transistor characteristics such as the threshold will change depending on whether these states have trapped charge or not. As miniaturization progresses, the relative effect of a single trap increases. Thus, when the drain current I_D is observed for the same transistor under the same bias conditions, a steplike change in I_D as a function of time can be observed as shown in Fig. 7.16c. This is called random telegraph noise (RTN). The time constant from when charge becomes trapped in an interface state until it is released is widely distributed from several ns to several s. Therefore, the spectrum of the current waveform in Fig. 7.16c will also be widely distributed.

RTN does not have a large effect up to 45 nm technology, but from 20 nm and beyond, the effects of RTN are expected to be as great as, if not greater than, the degradation due to NBTI, and even SRAM cells are thought to malfunction. This problem must be approached from both the device side and the circuit design side.

7.3.7 *Simulation of Degradation Prediction*

The values of variation in fabrication are determined at fabrication time and do not change thereafter, so these are predicted with Monte Carlo simulations. Soft error events occur instantaneously and can be simulated with current sources. RTN varies

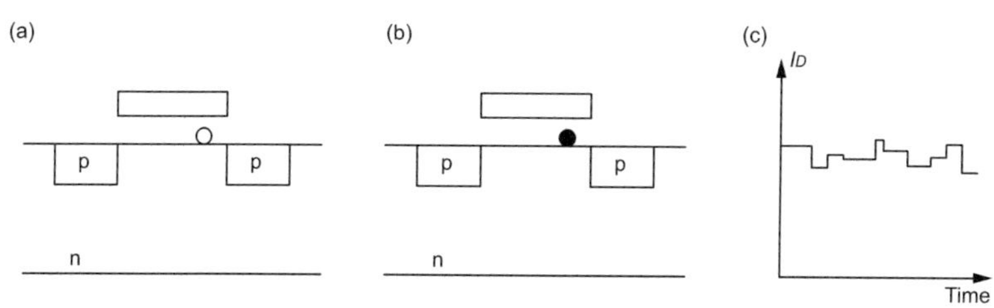

Fig. 7.16 Random telegraph noise

constantly in a random fashion and can be treated as noise power, along with shot noise and thermal noise. Degradations due to electromigration can be mitigated by designing interconnect that does not suffer degradation and does not need to be dealt with at circuit design time.

With regard to HCI and NBTI/PBTI, device characteristics not only change very slowly with time, but whether degradation or recovery occurs depends on the bias conditions. Therefore, the degradations must be simulated with SPICE. A model called MOS Reliability Analysis (MOSRA) is widely used for the degradation models, and most SPICE simulators include MOSRA models along with transistor models. By setting parameter values, the degree of degradation and recovery based on circuit operation can be predicted. Simulating only hot carriers, or only NBTI/PBTI, or both hot carriers and NBTI/PBTI are all possible. For example, in HSPICE, the simulation could be as follows:

```
*--------------------- Circuit Definition ---------------------
.OPTION POST=2 POST_VERSION=2001
.OPTION ACCURATE

.OPTION PROBE
.PROBE V(OUT) V(IN) V(n1) V(n2)
.param mvdd = 1.8
.TEMP = 27

.TRAN 0.1p 2000n
VIN IN 0 pwl(0n 0, 10n 0, 10.05n mvdd, 1990n mvdd, 1990.05n 0)
VVDD VDD 0 DC mvdd
VGND GND 0 DC 0

m0 n1 in gnd gnd n L=180e-9 W=700e-9
m1 n1 in vdd vdd p L=180e-9 W=2e-6
m2 n2 n1 gnd gnd n L=180e-9 W=700e-9
m3 n2 n1 vdd vdd p L=180e-9 W=2e-6
m4 out n2 gnd gnd n L=180e-9 W=700e-9
m5 out n2 vdd vdd p L=180e-9 W=2e-6

*---------------------- MOSRA model --------------------------
.model p_m mosra                * model definition
+ level=1
+ tit0=2e-4                     * delta-Vth is proportional to tit0
+ titfd=7.5e-10                 * refer to the manual for
+ tittd=-1.45e-20               * the parameter details
+ tn=0.25
+ ttd0=1 tdcd=-2.8              * these two parameters relates to
                                * the recovery

* model replacement
.appendmodel p_m mosra p pmos

.mosra
+ reltotaltime=3.15e+7          * stress time
+ relmode=2                     * 0:HCI&BTI, 1:HCI, 2:BTI
+ simmode=3                     * 0:before stress, 1:after stress,
```

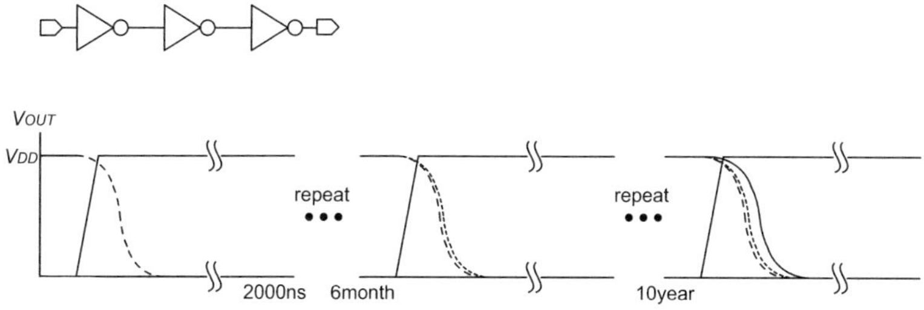

Fig. 7.17 Degradation prediction simulation with MOSRA

```
+                                * 2:before and after stress
+                                * 3:continuous
+ relstep=6.3e+6                 * time step for the observation
*+ AgingPeriod=6.3e+7            * AW is under stress among AP
*+ AgingWidth=3.15e+7
+ aginginst="m3"                 * specify the stressed instance.
                                 * all instances are stressed
                                 * if not specified

.MODEL N NMOS
+ LEVEL = 53
+ VERSION = 3.2


.MODEL P PMOS
+ LEVEL = 53
+ VERSION = 3.2

.END
*---------------------------------------------------------------
*mtest.tr0           simulation results without stress
*mtest.tr0@ra.grp    generated tr0 file lists
*mtest.tr0@ra=xxx    simulation results after xxx[sec]
*                    cscope cannot read the files with
*                    multiple dots
*mtest.radeg0        necessary file for continuous stress.
*                    dVth is written.
*---------------------------------------------------------------
```

In this example, a 2000 ns simulation is conducted. By applying this condition repeatedly, the characteristics slowly degrade, and the state of degradation every 6.3×10^6 s = 6 months is simulated (Fig. 7.17).

Chapter 8
Measurement Devices

After the chip has arrived, it is time for measurements. What? Sampling oscilloscopes and real-time oscilloscopes? Aren't oscilloscopes just oscilloscopes? Huh? We can't make measurements because there is no trigger signal output? Oh no... To avoid such mistakes, circuit designers must also be aware about measurement devices.

8.1 Sources of Signals to the Chip

8.1.1 Power Supply

First of all we have the power supply. As shown in Fig. 8.1a, there are two output terminals, and the specified constant voltage is continuously output regardless of the connected impedance and therefore regardless of the output current value or frequency. Usually there is a current limiter functionality, and when the current is pushed to above the threshold setting, an error occurs and the output becomes an open circuit or the voltage value decreases. For example, for a 1 V output and a setting of 100 mA maximum output current, connecting a 1 Ω resistance will cause an error. This functionality is to prevent a large current running through and burning a chip in case an unexpected supply-ground short exists. A setting of about 300 mA should be sufficient, although this depends on what is being measured. Also, of the two terminals, one is often the ground potential in normal usage situations.

The internal circuit is shown in Fig. 8.1b, for example. Usually, the D/A converter that generates the specified voltage is unable to supply a large current, so a unity gain amplifier structure is utilized for a large output current. However, with such a structure, the current value with frequency components higher than the bandwidth of the unity gain amplifier is unable to keep up. A general power supply has a low-pass

T. Nakura, *Essential Knowledge for Transistor-Level LSI Circuit Design*,
DOI 10.1007/978-981-10-0424-7_8

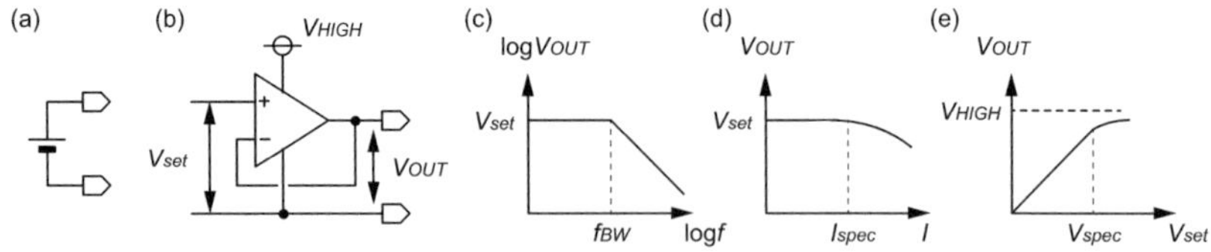

Fig. 8.1 (**a**) Power supply, (**b**) internal circuitry, and (**c**) (**d**) (**e**) power supply voltage characteristics

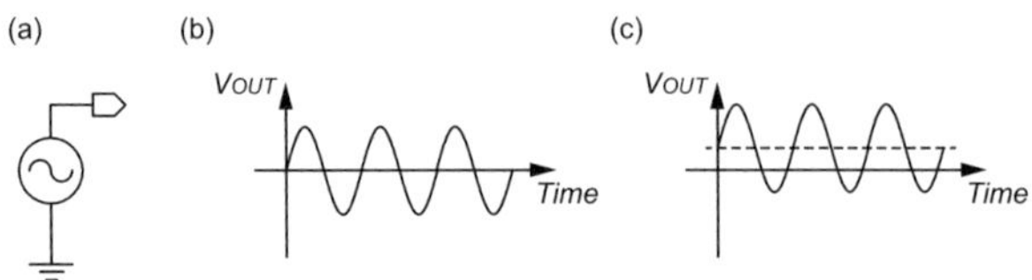

Fig. 8.2 (**a**) Symbol for SG, (**b**) output waveform, (**c**) added DC bias

frequency characteristic as shown in Fig. 8.1c. Also, if the current limit is removed and the load resistance is lowered, a large current will be output, but as shown in Fig. 8.1d when the current increases beyond a certain amount, the output voltage decreases. A rated current I_{spec} is specified as a specification for the supply. If the voltage setting is increased even while the current is suppressed to below the rated level, voltages beyond a certain value will not be output. This is the rated voltage V_{spec}.

Because these power supplies utilize electric circuits at the output stage, they contain some small amount of noise. If this noise bothers you, a dry cell battery is (supposedly) a low-noise supply, although this raises other concerns such as the inability to set voltages, the rated current, and the electrical capacitance.

8.1.2 Signal Generators

Signal generators (SG) are circuits that create signals, and they output very clean sine waves. In general, the symbol representing these is as in Fig. 8.2a, and a signal is output to a transmission line with one terminal connected to the ground. As shown in Fig. 8.2b, the output voltage is usually centered around zero. Devices with the capability to add an optional DC voltage exist, as in Fig. 8.2c, but it is more common to adjust the DC voltage by utilizing an external bias tee (introduced in Chap. 9).

The frequency and the amplitude can be changed in the settings. The frequency is simply changed by specifying the frequency value in Hz, but the amplitude can be set in dBm, peak-to-peak voltage V_{pp}, or effective voltage V_{eff}. dBm is the measure of power with respect to 1 mW for a 50 Ω termination, in dB. 0 dBm means 1 mW,

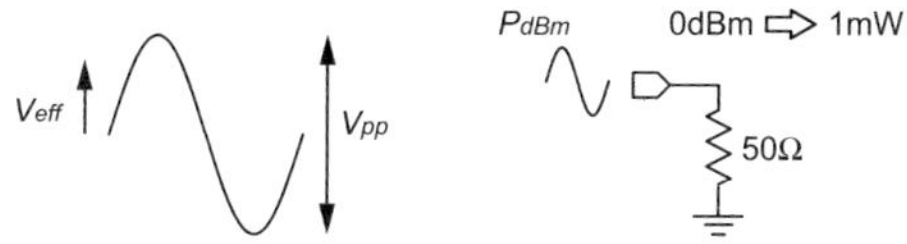

Fig. 8.3 dBm

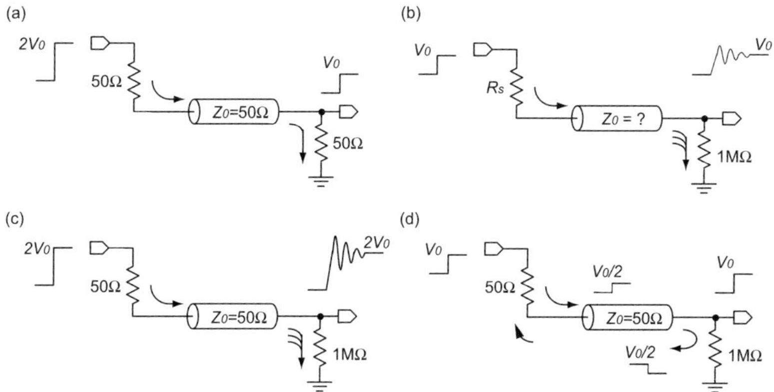

Fig. 8.4 Impedance and the output voltage

and since $P = V_{\text{eff}}^2/R$, $V_{\text{eff}} = \sqrt{PR} = \sqrt{1 \times 10^{-3} \times 50} = 0.2236\,[\text{V}]$. In general, the relationship between dBm power and the voltage is:

$$P_{\text{dBm}} = 10\log_{10}\frac{V_{\text{eff}}^2/R_{50}}{1\,\text{mW}} \quad \Leftrightarrow \quad V_{\text{eff}}^2 = 10^{\frac{P_{\text{dBm}}}{10}} \times R_{50} \times 1\,\text{mW}\,. \tag{8.1}$$

In addition, the effective value (RMS) of the voltage V_{eff} is $V = \sqrt{2}V_{\text{eff}}\sin(\omega t)$ and $V_{pp} = 2\sqrt{2}V_{\text{eff}}$ as shown in Fig. 8.3.

The output amplitude is adjustable from about -100 dBm (V_{pp} ~6 uV) to 15 dBm (V_{pp} ~3.5 V), and in general, the jitter is around several hundred fs and the higher harmonics are below -30 dBc (relative to the wanted frequency strength, the strength of the third harmonic is 1/1000). Also, some come with the option of AM/FM/PM.

Here, we will discuss output power, characteristic impedance, and termination resistance again. With any device that outputs signals (not just signal generators), the correct output voltage cannot be produced without an assumption about the input impedance of the input side. When building a measurement environment with 50 Ω as in Fig. 8.4a, V_0 can be input at the receiving end by having an output impedance of 50 Ω and outputting an internal voltage of $2V_0$. Having an output voltage of V_0 refers to this state. When 1 MΩ is expected at the receive terminal, an internal voltage equal to the output voltage is utilized to output the waveform as in

Fig. 8.4b. However, reflections should be expected because transmission lines with a characteristic impedance of 1 MΩ are rarely used.

A general GHz-class high-speed measurement instrument assumes the 50 Ω environment of Fig. 8.4a, and many MHz-class measurement instruments assume the 1 MΩ environment of Fig. 8.4b. The high voltage at the receiver would not rise all the way to the specified voltage when receiving a signal with 50 Ω from a signal generator that assumes the situation in Fig. 8.4b. Similarly, receiving a signal with 1 MΩ from a signal generator that assumes the situation in Fig. 8.4a would inject double the specified voltage V_0, as shown in Fig. 8.4c, and can potentially break the chip. It is necessary to confirm the type of instrument before using it and also to select the correct IO buffers at the design stage that match the measurement equipment. Also, some instruments allow a selection between 50 Ω termination or 1 MΩ termination at the receiving end, so a choice can be made accordingly. With this type of device, the output impedance is 50 Ω even when the 1 MΩ termination option is selected, as shown in Fig. 8.4d. Initially a voltage of $V_0/2$ is injected in to the transmission line. The voltage at the receiving terminal becomes V_0 due to the open reflection with reflection coefficient $\Gamma_L = 1$. The transmitting terminal has a matched termination, and the reflection is absorbed with reflection coefficient $\Gamma_s = 0$.

8.1.3 Pulse Pattern Generators

Pulse pattern generators have several data outputs for ONE/ZERO, as well as outputs of CLK synchronized to the data and 8× or 1024× slower CLKs for trigger signals as shown in Fig. 8.5. The data is programmable and can be specified to be output once, a certain number of times, or infinitely many times. The output frequency is either specified as a number in Hz or is synchronized to an external trigger input. The output amplitude can also be specified, and the termination resistance of the receiver (50 Ω/1 MΩ) often needs to be specified as well. When

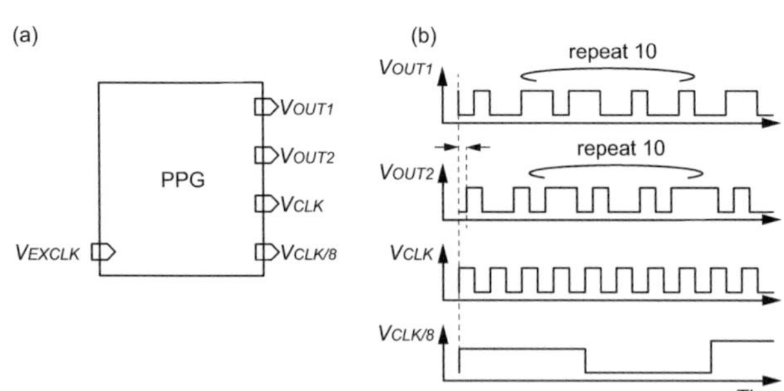

Fig. 8.5 (**a**) PPG, (**b**) output waveforms

multiple streams of data are output, the time delay between them is in some cases also programmable. In short, these are instruments for freely generating digital data.

8.2 Observers of Signals from the Chip

8.2.1 Sampling Oscilloscopes

8.2.1.1 The Principles

An oscilloscope is generally used to observe voltage changes over time. In a sampling oscilloscope as shown in Fig. 8.6a, the data waveform to be observed is assumed to be a periodic waveform, and a trigger signal synchronous to the period is necessary. As a trigger signal is input, a sample and hold signal *SH* delayed by $k\Delta t$ with respect to the trigger is generated internally in the oscilloscope. k is incremented by 1 every time a trigger is input. The voltage of the data waveform at the moment the *SH* pulse is generated is stored, and the input waveform is synthesized by drawing the stored values Δt apart.

8.2.1.2 Trigger

As can be seen from the principles, the trigger signal period must be an integer multiple of the data signal repeat period: $T_{\text{trigger}} = NT_{\text{repeat}}$. If a trigger signal with a period shorter than the period of the data signal is used, the sampling of the data waveform would be as the timing shown in Fig. 8.6c, and the synthesized waveform would become as shown in Fig. 8.6d.

8.2.1.3 The Sampling Period, Equivalent Time Sampling, and Temporal Resolution

The smaller the shift amount Δt for the *SH* signal is, the denser the sampling points are for the synthesized waveform, and more accurate waveforms can be obtained. This is the sampling frequency, where $f_{\text{sample}} = 1/\Delta t$. For example, if the sampling frequency is 50 Gsample/s, $\Delta t = 20$ ps.

In an actual oscilloscope with 50 GS/s, the waveform is observed at an even finer time interval. This is called equivalent time sampling (ETS). As shown in Fig. 8.6e, the sampling signal is internally shifted by $\Delta t/N$ (in the Figure, $N = 4$), and N observations are overlapped to sample the data waveform at very fine time steps.

The temporal resolution refers to the smallest time step of the synthesized waveform and is determined not by the sampling interval but rather the shift amount of the equivalent time sampling, $\Delta t/N$.

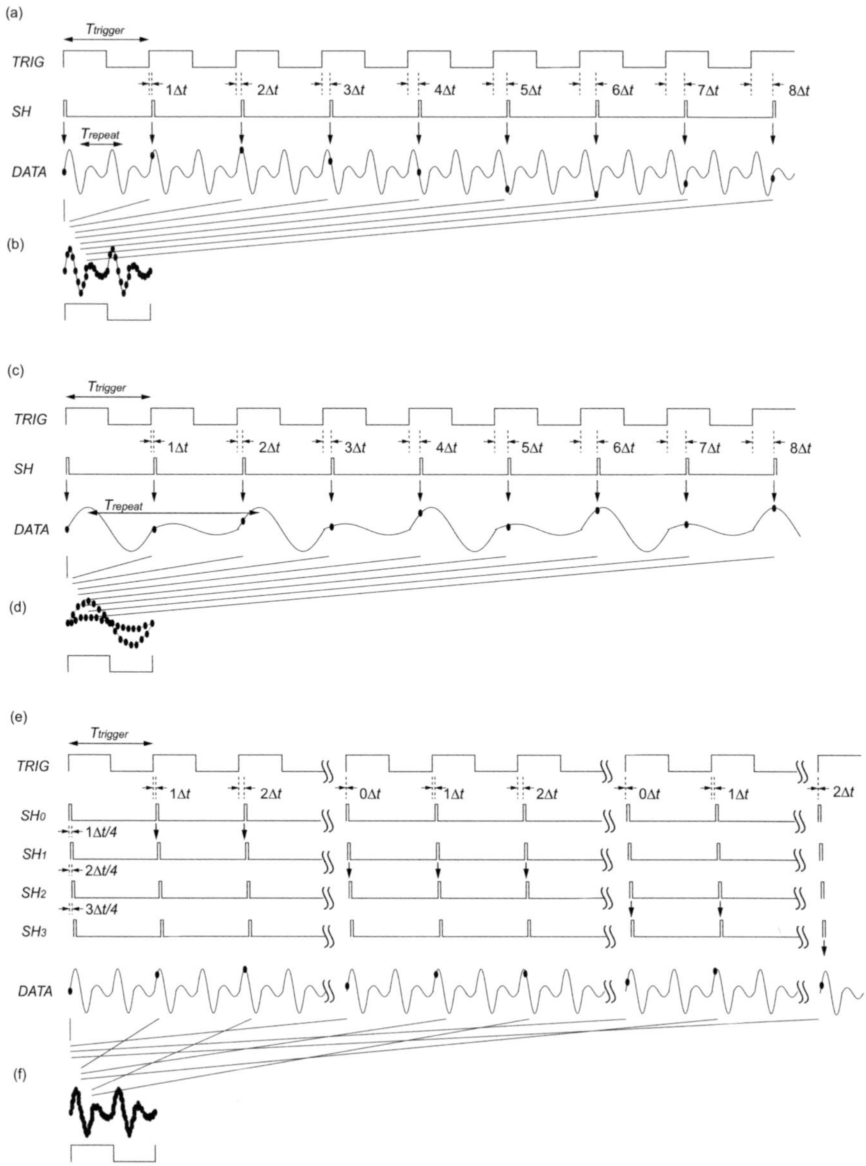

Fig. 8.6 Waveform sampling and synthesis of a sampling oscilloscope

8.2.1.4 Block Diagram

The internals of a sampling oscilloscope might look like, for example, Fig. 8.7. The sampling timing generator contains a delay chain with a delay of Δt in each stage to generate the sampling timing, as well as a DELAY block to generate a delay of $\Delta t/k$

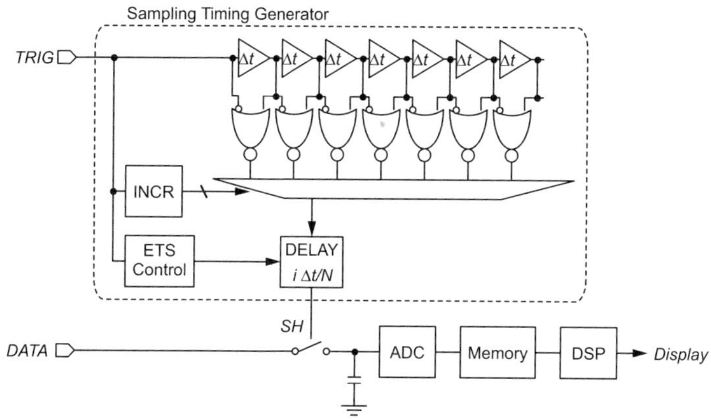

Fig. 8.7 The internal circuitry of a sampling oscilloscope

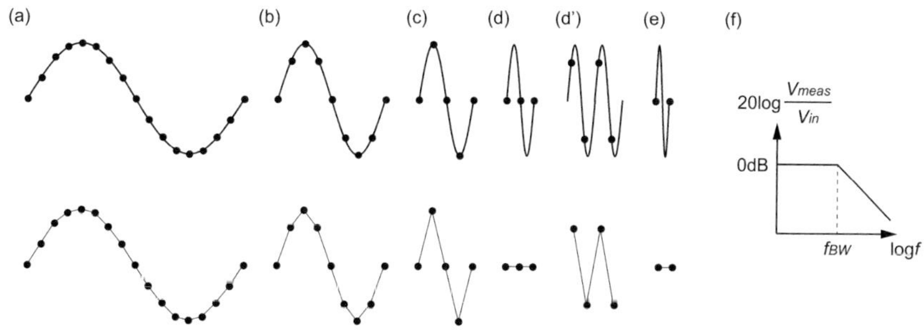

Fig. 8.8 Bandwidth

for ETS. The voltage obtained from sampling and holding the input signal with the *SH* signal is converted to digital data through an A/D converter circuit. The data is then stored in memory, processed with DSP, and finally displayed on the screen.

8.2.1.5 Bandwidth

The bandwidth indicates the highest frequency that can be sampled when sampling the data signal with the *SH* signal. In Fig. 8.8a–e, the top indicates the input waveform and the sampling points, and the bottom indicates the waveform generated by connecting the sampled points. As can be seen from this figure, it is not possible to sample signals which change in shorter time intervals than the pulse width of the *SH* signal. The oscilloscope bandwidth f_{BW} is the point where the synthesized waveform amplitude becomes $1/\sqrt{2}$ of the original amplitude, when sweeping the frequency of the input sinusoid as indicated in Fig. 8.8f.

Theoretically, the bandwidth is determined by the sampling period, where the relationship between the bandwidth f_{BW} and sampling period is $f_{BW} = 1/(2\Delta t/N)$ from the Nyquist sampling theorem. For example, in Fig. 8.8a–c, the frequency of the original waveform is lower than the Nyquist frequency, which is half of the sampling frequency. Assuming that there is no frequency content at frequencies higher than the Nyquist frequency, the original top waveforms can be reconstructed from the bottom sampling results. This assumption does not hold in (d)–(e), and the original signal cannot be reconstructed.

In reality, the actual bandwidth is degraded from the theoretical value determined by the sampling period, due to factors such as the bandwidth of the probing head and the characteristics of the sample and hold circuit within the oscilloscope.

Thus, it is important to keep in mind that even if the oscilloscope has high temporal resolution with ETS, the bandwidth is not that high and is actually determined by the bandwidth of the probing head and the characteristics of the internal sample and hold circuitry.

8.2.1.6 Input Impedance

The quickest way to master the use of oscilloscopes after understanding the principles is to actually try out measurements and fiddle with settings to observe how the waveforms are displayed, but I will comment on the input impedance here.

As shown in Fig. 8.9, the LSI and oscilloscope are connected through a transmission line. Let the output impedance of the LSI be R_s and the input impedance of the oscilloscope be R_{IN}. In general, the input impedance of the oscilloscope can be switched between 50 Ω and 1 MΩ. When dealing with high-speed signals, a transmission line with $Z_0 = 50\ \Omega$ and $R_{IN} = 50\ \Omega$ is used to prevent signal reflections. Reflections can be suppressed by consistently using 50 Ω, but the observed voltage level will not be V_0 but rather $V_0 R_{IN}/(R_s + R_{IN})$. The observed voltage level will be V_0 if the input impedance of the oscilloscope is set to $R_{IN} = 1\ \mathrm{M}\Omega$, but only slow signals can be observed this way because reflections will occur due to termination impedance mismatch of the transmission line. The transmission line characteristic impedance and the input impedance of the oscilloscope should be decided based on reflections, the signal speed, and voltage levels. Also, when designing (selecting) the IO during chip design, measurement should be taken into account. For example,

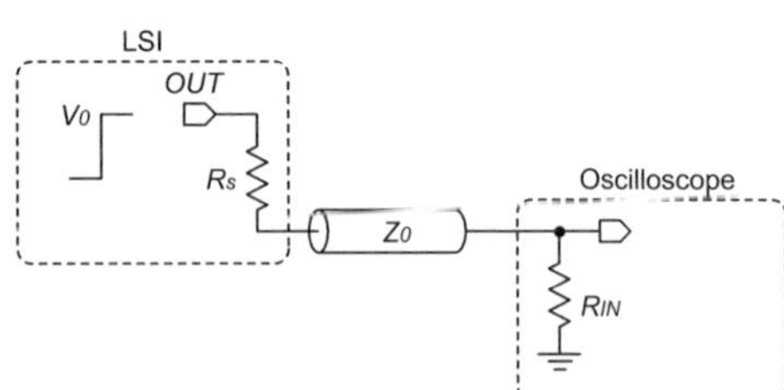

Fig. 8.9 Input impedance of an oscilloscope

"this signal is fast and will be captured with a 50 Ω system; so to make the output impedance of the output buffer R_s small, I should use a transistor with large W."

8.2.1.7 Eye Patterns

In the process of making LSI for communications, the observation of the eye pattern becomes necessary, and this is measured using an oscilloscope. As shown in Fig. 8.10, a digital signal waveform of ONE/ZERO that randomly changes synchronously with CLK is repeatedly superimposed using the trigger signal. If the waveform is clean as shown in Fig. 8.10a, the result of the overlaying will be the shape shown on the right. This is called the eye pattern or eye diagram, from its similarity to the shape of a human eye. In this text, we will call it the eye pattern. When deviations in the time axis occur due to jitter as shown in Fig. 8.10b, the eye pattern is blurred horizontally. When deviations occur in the voltage axis as in Fig. 8.10c, the eye is blurred vertically. When random noise sits on top of the signal as shown in Fig. 8.10d, the lines in the eye pattern become thicker. With the functionality to portray the frequency of the overlay with shading, an eye pattern as in Fig. 8.10e can be obtained. Eye patterns are widely utilized as a simple method to display the received signal integrity of a random ONE/ZERO digital signal over a

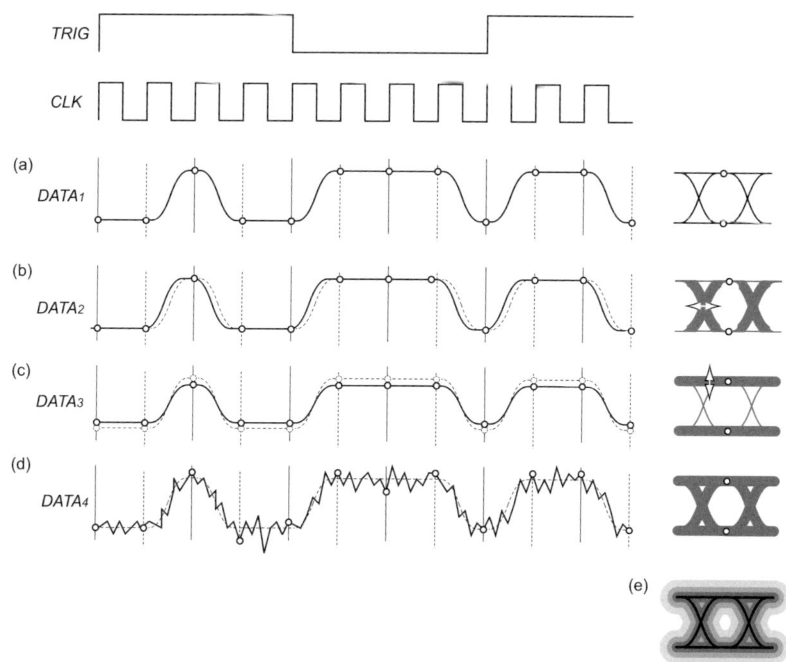

Fig. 8.10 Measuring the eye pattern

long period of time, and many oscilloscopes come with an eye pattern measurement mode. Well, all it takes is to do a regular waveform measurement and repeatedly overlay the signal for display. Fancy oscilloscopes will not only display the eye pattern but also will automatically calculate the eye opening ratio or the estimated bit error rate.

8.2.1.8 Jitter Histograms

Jitter measurements are necessary when making phase-locked loops (PLLs), and this is also measured with an oscilloscope. As shown in Fig. 8.11a, a PLL is a circuit that, given a reference clock CLK_{REF} with frequency f_{ref} as input, outputs a clock CLK_{PLL} with an integer multiple of the input frequency, Nf_{ref}. The reference clock comes from a crystal oscillator and is a clean source, but the output clock of the PLL has noise (jitter). To determine the amount of jitter, the jitter histogram is measured. When CLK_{REF} is split and used as the PLL input as well as the trigger input for the oscilloscope and the PLL output is measured, the PLL can output a clock with small jitter as with CLK_{PLL0} in Fig. 8.11b, or the clock can have a large jitter as with CLK_{PLL1}. Just as with the eye pattern, the PLL output can be repeatedly superimposed to generate something like shown in Fig. 8.11c, and the frequency of the timing at which the signal crosses some constant voltage (usually, half of the output amplitude) is shown as a histogram. Many oscilloscopes come with a jitter measurement mode. Usually, the jitter distribution is a normal distribution, and the standard deviation indicates the amount of jitter. Again, all it takes to do this is to

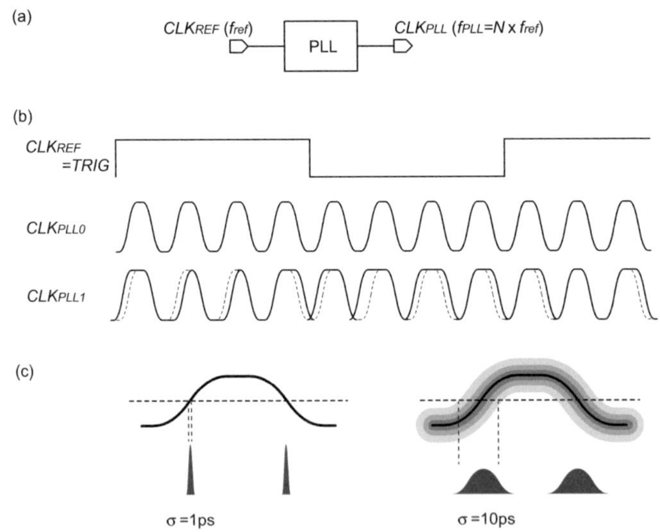

Fig. 8.11 Jitter histogram measurement

do a normal waveform measurement and overlay the signal for display and also show the number of times the signal crosses the specified voltage as a histogram. Normally, the standard deviation and the peak-to-peak values are also automatically displayed.

8.2.2 *Real-Time Oscilloscopes*

A sampling oscilloscope was only able to observe periodic waveforms that were synchronized to the trigger signal, but a real-time oscilloscope can observe nonperiodic waveforms (of course, periodic waveforms can be observed as well). As shown in Fig. 8.12a, data is captured at constant intervals of Δt from the first rising edge of the trigger signal until the memory is full. In some cases, data can be captured from some certain time before the trigger rising edge.

A basic block diagram is shown in Fig. 8.12b. Pulses at time interval Δt are output from the periodic pulse generator circuit. The SR latch is reset beforehand. By setting the SR latch with the trigger, the periodic pulse signal is output to the sample and hold circuit, and the input data is obtained and observed.

However, with this method, the sample and hold circuit as well as the A/D converter circuit must operate at Δt intervals, and therefore, the sampling period cannot be made shorter. Thus, as shown in Fig. 8.12c, d, N blocks can be interleaved so that each sample and hold circuit and A/D converter circuit only needs to operate at intervals of $N\Delta t$, allowing for a shorter sampling period. The case for $N = 4$ is shown in Fig. 8.12c, d.

The equivalent time sampling concept, which assumes periodic waveforms, has no meaning with real-time oscilloscopes. However, real-time oscilloscopes generally have a real-time measurement mode and a sampling measurement mode that assumes periodic waveforms, and with the sampling mode, a high resolution can be obtained with ETS.

8.2.2.1 Jitter Spectrum

The eye pattern and the jitter histogram that were observed with the sampling oscilloscope can also be observed (obviously) by utilizing a real-time oscilloscope. In addition, a real-time oscilloscope can measure the change in jitter with time, which could not be measured with a sampling oscilloscope. For example, if the waveform in Fig. 8.11 is observed, a sampling oscilloscope could only measure the histogram. However, a real-time oscilloscope can observe that relative to the ideal period, the rising edge jitter changes as -1, $+1$, $+1$, 0, -1, -2, -1, 0, 0, and 0 ps with time. By applying the Fourier transform to this, the jitter spectrum, which contains the frequency contents of the jitter, can be calculated.

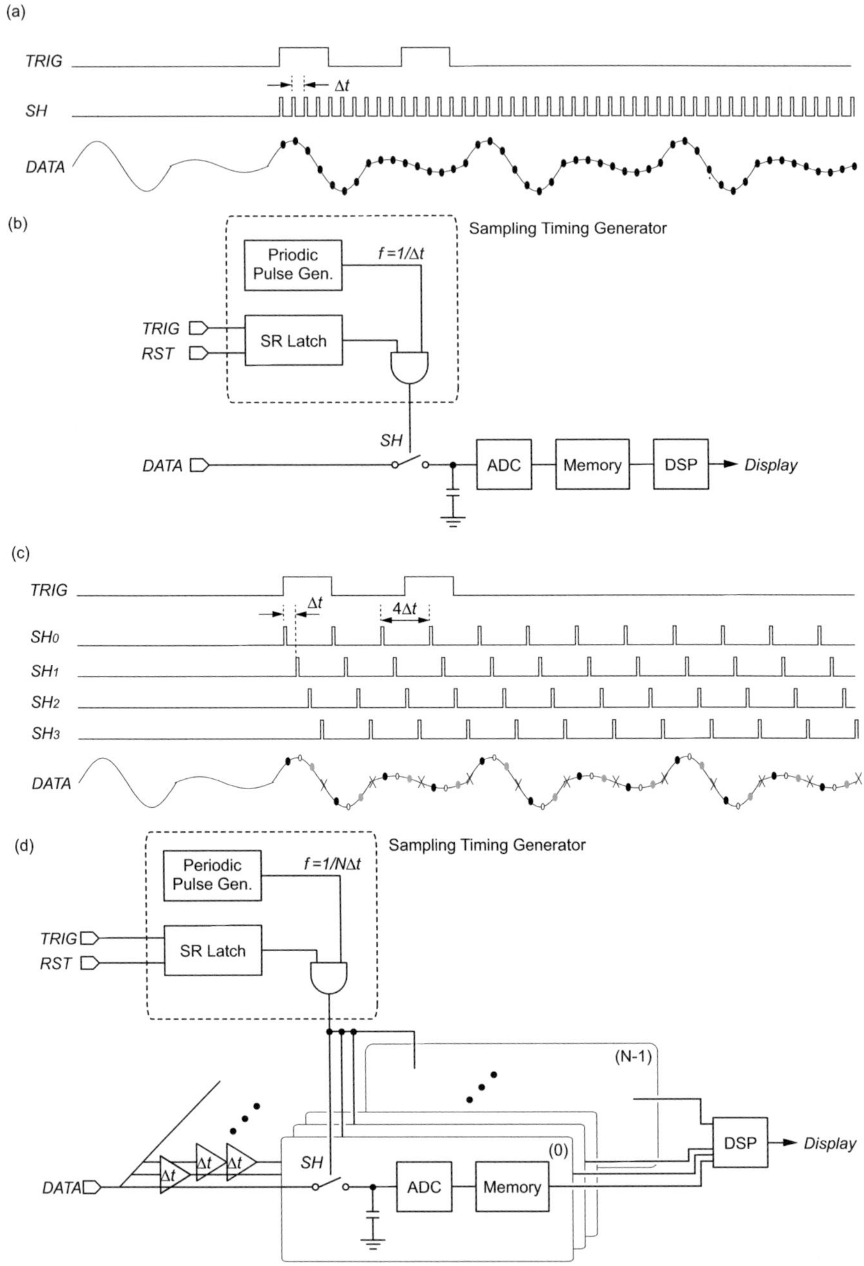

Fig. 8.12 Measurement waveform and block diagram of a real-time oscilloscope

8.2.3 *Spectrum Analyzers*

Spectrum analyzers display the frequency content of a signal.

8.2.3.1 The Principles

The frequency contents of a signal, as shown in Fig. 8.13a, can be obtained by observing the variations of the signal voltage with time with a real-time oscilloscope and applying an FFT. However, because the measureable frequency bandwidth would be limited by the sampling period of the real-time oscilloscope, this method is rarely utilized.

Instead, a structure as in Fig. 8.13b is widely used. As background theory, when two sinusoids with different frequencies are multiplied together,

$$V_S(t) = A_S \cos(\omega_S t + \theta) \tag{8.2}$$

$$V_L(t) = A_L \cos(\omega_L t) \tag{8.3}$$

$$V_S(t) \times V_L(t) = \frac{A_S A_L}{2}[\cos\{(\omega_S + \omega_L)t + \theta\} + \cos\{(\omega_S - \omega_L)t + \theta\}] \tag{8.4}$$

and thus a term with the sum of frequencies and a term with the difference of frequencies are generated. In general, low-frequency components are easier to deal with in electronic circuits. Thus, a filter is used to extract the term with the frequency difference. If the input signal is V_S and the local oscillator (LO) signal is V_L, and if the V_L signal is completely known, then A_S can be calculated from the magnitude of the $\omega_S - \omega_L$ component ($A_S A_L/2$), and the frequency components of V_S are known.

In this example, we assumed that the only frequency component of V_S is at ω_S, but if V_S has a spread in frequency as shown in Fig. 8.13c, the frequency content of the output of the mixer will be as shown in the graph to the right, where the shape of the frequency spread of V_S is maintained and shifted to the left to lower frequencies by f_L (down-converted). Since sharp, narrow-band band-pass filters can be created at low frequencies, the frequency components contained in the original signal can be obtained by measuring the energy within that band. As shown in Fig. 8.13d, the amount of input signal down-conversion is changed by sweeping the LO frequency and keeping the band-pass filter constant, which is equivalent to sweeping the band-pass filter.

In practice, to realize a more precise spectrum measurement, it is common to execute the frequency conversion in two or three stages as shown in Fig. 8.13d.

8.2.3.2 The Resolution Bandwidth and the Noise Floor

The width of the band-pass filter of the last stage is called the resolution bandwidth (RBW), and this indicates the precision of the spectrum width. The spectrum

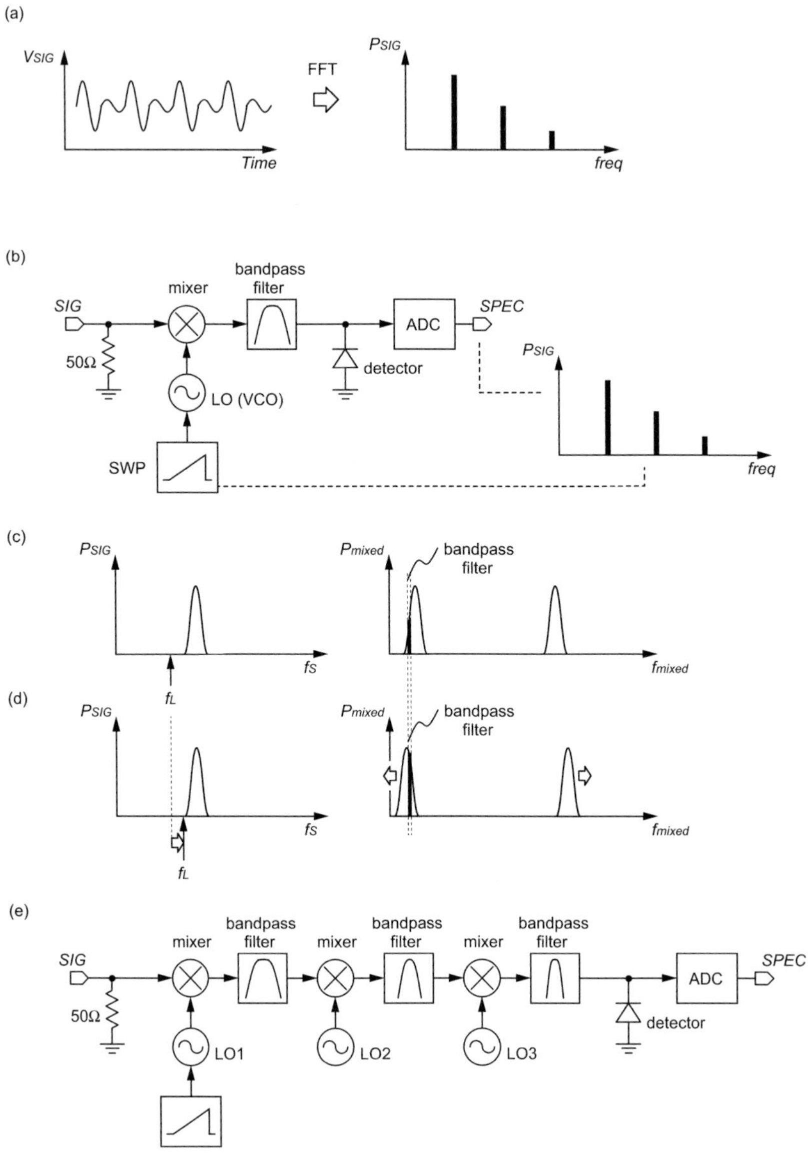

Fig. 8.13 Principles of a spectrum analyzer

analyzer displays the power contained in the input signal within this frequency width.

In general, resistance generates thermal noise. Here, thermal noise is the noise that is generated by the random heat vibrations of the electrons inside the resistive element. The noise voltage v_n that is generated within the resistive element at temperature T is:

$$v_n^2 = 4kTR\Delta f \tag{8.5}$$

Δf is the band of interest. The maximum noise energy input per 1 Hz from a noise source is:

$$P_n = \frac{v_n^2}{4R} / \Delta f = kT \tag{8.6}$$

which is a constant value regardless of the resistance value or the frequency. Here, k is the Boltzmann constant. The value at 300 K is:

$$P_n = 1.38 \times 10^{-23}\,[\mathrm{J/K}] \times 300\,[\mathrm{K}] = 4.14 \times 10^{-21}\,[\mathrm{W/Hz}] \tag{8.7}$$

which, expressed in dBm, is:

$$10 \log_{10} \left(\frac{4.14 \times 10^{-21}}{1 \times 10^{-3}} \right) = -174\,[\mathrm{dBm/Hz}]\,. \tag{8.8}$$

Thus, thermal noise of −174 dBm per 1 Hz is constantly being generated. Thus, the noise floor of a spectrum analyzer with an RBW of 1 MHz is:

$$P_n = -174\,[\mathrm{dBm/Hz}] + 10 \log_{10}(1 \times 10^6) = -174 + 60 = -114\,[\mathrm{dBm/Hz}]\,. \tag{8.9}$$

It is necessary to measure with the optimal RBW, by taking into account the purity of the signal frequencies as well as the signal levels.

8.3 Equipment with Both Signal Input and Output

8.3.1 *BERT*

A bit error rate tester (BERT) is used to test whether the ONE/ZERO digital values are correctly transmitted and received when the channel is unstable, for example, if there is noise or if there is a large loss. There is a close relationship to the eye patterns that we learned about in the previous section.

A digital pattern that actually is transmitted is a random mix of ONEs and ZEROs. A truely "random" signal is not suited for transmission tests due to its random nature, and pseudorandom bit streams (PRBS) are used instead. A pseudorandom signal is generated using circuits as shown in Fig. 8.14a–c. Usually, a default value of ONE is given to each FF. In (a), the 7 bit pattern of 1, 1, 1, 0,

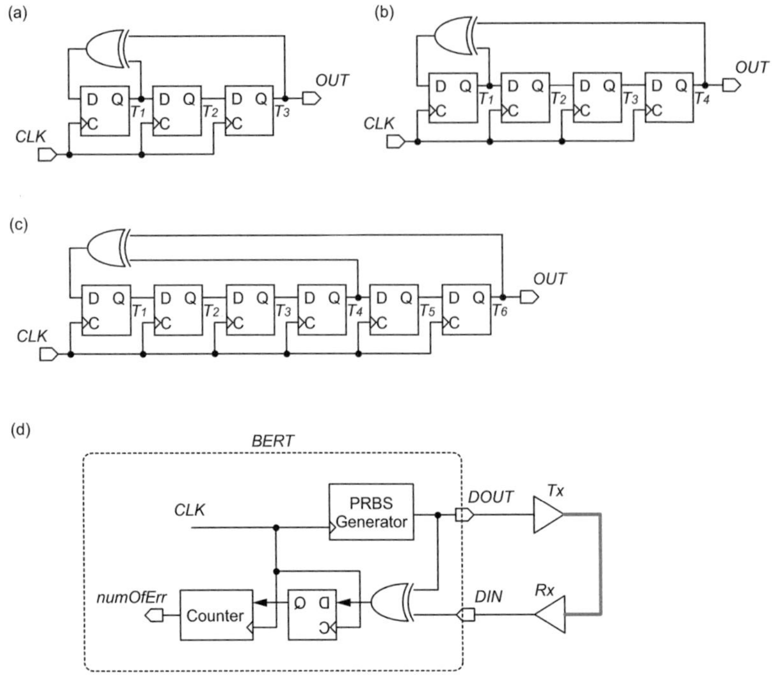

Fig. 8.14 PRBS generator circuit and BERT

1, 0, and 0 is output repeatedly. In general, with m shift registers, a pseudorandom repeating signal of $2^m - 1$ bits is generated, leading to periods in (a), (b), and (c) of $2^3 - 1 = 7, 2^4 - 1 = 15$, and $2^5 - 1 = 31$ bits, respectively. Also, the pattern contains m repeating ONEs and $m - 1$ repeating ZEROs. The number and location of signals to input to the XOR gate is determined by "a minimum shift algorithm generated by Galois field arithmetic," but the details are mathematics that circuit designers need not worry about. In the evaluation of LSI transmission, $m = 7, 15, 23, 31$ are often used.

In a BERT as shown in Fig. 8.14d, the characteristics of the transmitting and receiving circuitry as well as the transmission line can be measured by outputting a PRBS signal and counting the number of times a transmission error occurs (although not shown in the figure, it is also normal to have circuits that adjust the phase between the input signal *DIN* and the internal *CLK* or circuits that correct for *N* clock cycles of latency).

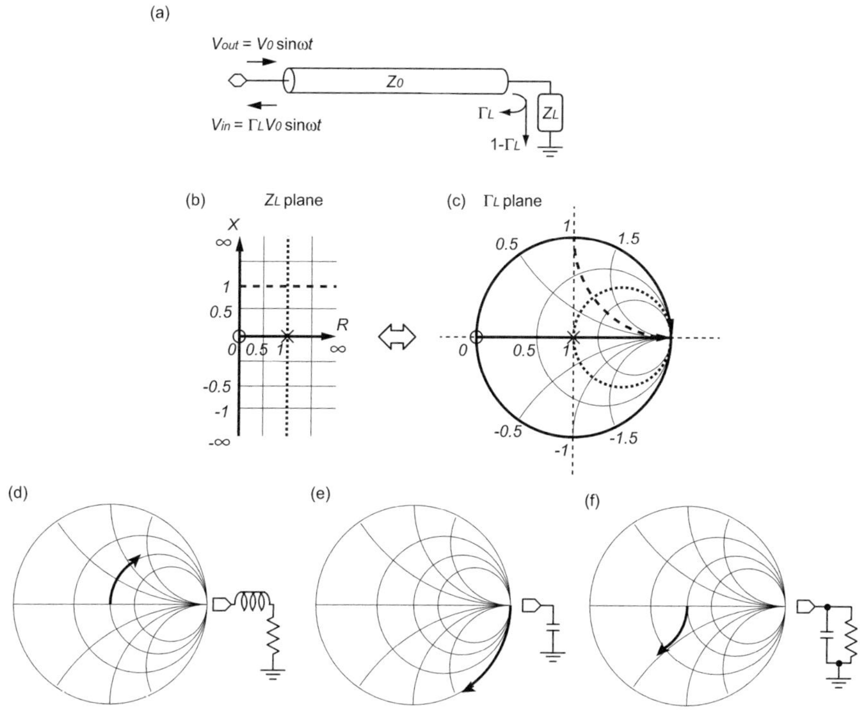

Fig. 8.15 Smith chart principles

8.3.2 *Network Analyzers*

8.3.2.1 Smith Charts

When a transmission line is in use, as shown in Fig. 8.15a, the signal sent from the transmitting end will reflect and come back. This reflectance is a function of the characteristic impedance Z_0 and the termination impedance Z_L, and we learned that:

$$\Gamma_L = \frac{Z_L - Z_0}{Z_L + Z_0} \tag{8.10}$$

in Sect. 5.1.3. If Z_0 is known and the reflectance Γ_L is also known, Z_L can be derived. That is, by outputting a signal of $V_0 \sin(\omega t)$ and measuring the amplitude and phase of the reflected wave, $Z_L(\omega)$ can be determined. By repeating this procedure while sweeping the frequency, $Z_L(\omega)$ can be known. From the frequency characteristics of Z_L, we can extract whether Z_L is resistive, capacitive, or inductive or how each of these is combined.

As long as the termination impedance does not contain any elements with amplification, the reflectance will never be greater than 1. Now, let:

$$\Gamma_L \ (= |\Gamma_L| e^{j\theta}) = u + jv \tag{8.11}$$

$$\hat{Z} = \frac{Z_L}{Z_0} = \hat{R} + j\hat{X} \tag{8.12}$$

and substitute into Eq. (8.10) and group terms to get:

$$\hat{R} = \frac{(1-u)^2 - v^2}{(1-u)^2 + v^2} \tag{8.13}$$

$$\hat{X} = \frac{2v}{(1-u)^2 + v^2} \tag{8.14}$$

From Eq. (8.13),

$$\left(u - \frac{\hat{R}}{\hat{R}+1}\right)^2 + v^2 = \left(\frac{1}{\hat{R}+1}\right)^2 \tag{8.15}$$

which means that the Γ_L for a constant $\hat{R}$ and a variable $\hat{X}$ represents a circle with its center at $\left(\frac{\hat{R}}{\hat{R}+1}, 0\right)$ and a radius of $\frac{1}{\hat{R}+1}$ and always passes through (1, 0). Similarly by Eq. (8.14):

$$(u-1)^2 + \left(v - \frac{1}{\hat{X}}\right)^2 = \left(\frac{1}{\hat{X}}\right)^2 \tag{8.16}$$

which means that the Γ_L for a constant $\hat{X}$ and a variable $\hat{R}$ represents a circle with its center at $\left(1, \frac{1}{\hat{X}}\right)$ and a radius of $|\frac{1}{\hat{X}}|$ and always passes through (1, 0). That is, Eq. (8.10) is a projection of the Z_L plane $\left(\hat{R}, \hat{X}\text{plane}\right)$ shown in Fig. 8.15b to the Γ_L plane shown in Fig. 8.15c. The graph in Fig. 8.15c is called a Smith chart. Also, the point for matched impedance where $Z_L = Z_0$ $\left(\hat{R} = 1, \hat{X} = 0\right)$ corresponds to the point for $\Gamma_L = (0, 0)$, which agrees with the fact that the magnitude of the reflection with matched termination is zero.

By plotting the measured results of the reflections on a Smith chart as the frequency is changed, the behavior of Z_L is known. Plot examples for some termination circuits and the reflectances on Smith charts are shown in Fig. 8.15d–f.

8.3.2.2 S Parameters

A forward wave V_f and backward wave V_b exist in a transmission line. We now define:

$$a \equiv V_f/\sqrt{Z_0} = I_f\sqrt{Z_0} \tag{8.17}$$

$$b \equiv V_b/\sqrt{Z_0} = I_b\sqrt{Z_0} \tag{8.18}$$

The square of the absolute values of these values are:

$$|a|^2 = |V_f|^2/Z_0 = |I_f|^2 Z_0 \tag{8.19}$$

$$|b|^2 = |V_b|^2/Z_0 = |I_b|^2 Z_0 \tag{8.20}$$

which are the powers of the forward and backward waves. Also, the reflection coefficient is:

$$\Gamma = \frac{V_b}{V_f} = \frac{b\sqrt{Z_0}}{a\sqrt{Z_0}} = \frac{b}{a} \tag{8.21}$$

In addition, at a certain point, the voltage is $V = V_f + V_b$ and the current is $I = I_f - I_b$.

Similarly, when a 4-terminal network is considered as in Fig. 8.16, the equations relating a and b:

$$\begin{pmatrix} b_1 \\ b_2 \end{pmatrix} = \begin{pmatrix} S_{11} & S_{12} \\ S_{21} & S_{22} \end{pmatrix} \begin{pmatrix} a_1 \\ a_2 \end{pmatrix} \tag{8.22}$$

expressed in this way are called S parameters (which might not mean anything at all right now, so I will go into more detail).

From Eqs. (8.19) and (8.20), we see that $|a|^2$ and $|b|^2$ represent power, and a and b are proportional to voltage as well as current. Equation (8.21) can be thought to come from $b = \Gamma a$, and Eq. (8.22) is an extension of the reflectance as well as transmittance to a multi-port network. That is:

$$S_{11} = \left.\frac{b_1}{a_1}\right|_{a_2=0} = \left.\frac{V_{b1}}{V_{f1}}\right|_{a_2=0} \tag{8.23}$$

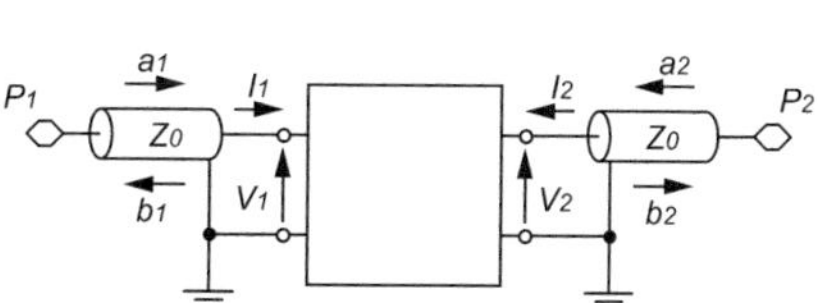

Fig. 8.16 S parameters

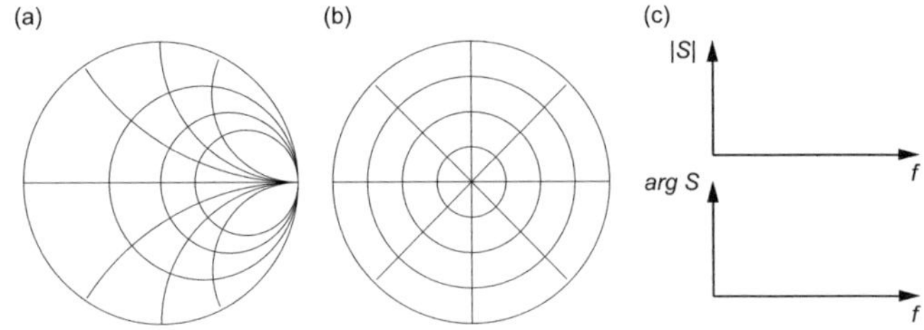

Fig. 8.17 S parameter plots

where S_{11} represents the reflectance of port 1 when node P_2 has matched termination (with the impedance of Z_0 added). Similarly, S_{22} represents the reflectance of port 2 when node P_1 has matched termination. On the other hand,

$$S_{21} = \left.\frac{b_2}{a_1}\right|_{a_2=0} = \left.\frac{V_{b2}}{V_{f1}}\right|_{a_2=0} \tag{8.24}$$

where S_{21} represents the transmittance from port 1 to port 2 when node P_2 has matched termination. Similarly, S_{12} represents the transmittance (amplification factor) from port 2 to port 1 when node P_1 has matched termination.

A network analyzer measures the dependence of these reflectance and transmittance (gain) parameters. In general, S_{11} and S_{22} are represented with a Smith chart as in Fig. 8.17a, and S_{21} and S_{12} are represented with a polar chart as in Fig. 8.17b. Both are simply plotting S on the complex plane, and only the underlying grid pattern is different. For some applications, S can be plotted in a Bode plot with magnitude and phase separated, as in Fig. 8.17c.

In an RLC circuit, $S_{21} = S_{12}$. Also, $|S_{11}|^2 + |S_{21}|^2 = 1$, and $|S_{22}|^2 + |S_{12}|^2 = 1$. Of course, these relationships break down when a transistor is inserted, and S_{21} can become greater than 1 and go outside of the unit circle.

A network analyzer is used to measure, for example, whether termination is matched across a wide frequency range (whether $|S_{11}|$ is small enough) or whether the reflectance is small and gain is large only at a particular target frequency. Also, if the frequency response of the impedance is known, the characteristics of Z_L are completely known, which means that with a bit of mathematics such as convolution, the time response can also be calculated.

8.3.2.3 Calibration

Normally, the network analyzer itself and the target for measurement are physically separated, and they are connected through transmission lines. The network analyzer measures only the waveform output from its own port and the waveform input to its own port, which means that it measures the S parameters of the transmission line + the target for measurement. If the transmission line has ideal characteristics and the length is known, the S parameters of the measurement target can be

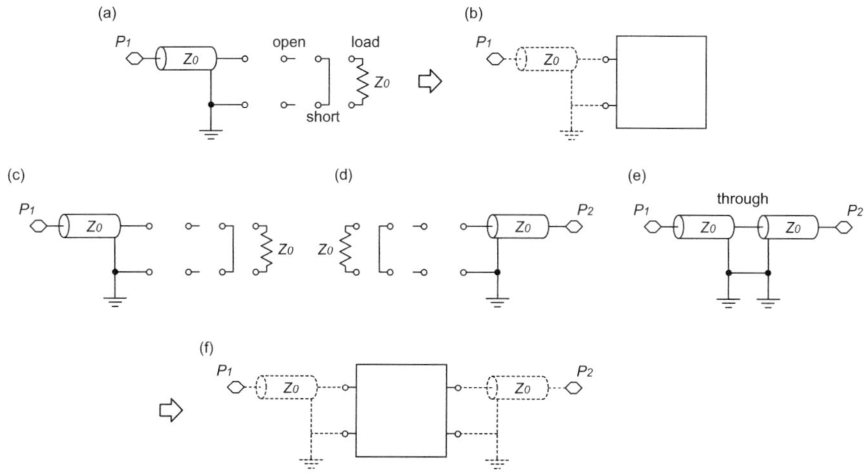

Fig. 8.18 Calibration for the measurement of S parameters

calculated from the measurement results of the transmission line + the measurement target, but an actual transmission line is nonideal and has attenuation and frequency-dependent characteristics. Therefore, the characteristics of the transmission line only are first measured, after which the characteristics of the transmission line + the measurement target are measured and the S parameters of only the target for measurement can be calculated. This is called calibration.

In the calibration for a 1-port S_{11} measurement, the reflected waves are first measured after connecting an open, a short, and a load (also called a match) at the end of the transmission line, as shown in Fig. 8.18a. Then, as shown in Fig. 8.18b, the characteristics without the effects of the transmission line are calculated by connecting the measurement target and measuring the S parameters. In the calibration for a 2-port $S_{11}, S_{21}, S_{12}, S_{22}$ measurement, an open, a short, and a load are connected to the ends of the transmission lines for port 1 as well as port 2, and the reflected waves are measured, as shown in Fig. 8.18c, d. Then, as shown in Fig. 8.18e, the transmission lines are connected in a "through" fashion, and the reflected and transmitted waves are measured. Finally, as shown in Fig. 8.18f, the measurement target is connected to measure the reflection from port 1 to port 1, the transmission from port 1 to port 2, the transmission from port 2 to port 1, and the reflection from port 2 to port 2, which allows the calculation of $S_{11}, S_{21}, S_{12}, S_{22}$ with the effects of the transmission lines removed.

8.3.3 Logic Analyzers

Logic analyzers determine the ONE/ZERO of incoming data. The basic structure is as shown in Fig. 8.19a: the ONE/ZERO of the input data is sequentially written

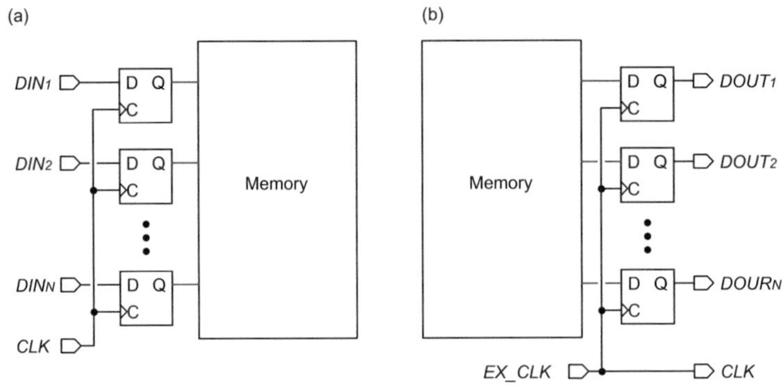

Fig. 8.19 A logic analyzer and a logic generator

into memory and displayed as necessary. An actual logic analyzer is slightly more complicated and allows for things such as the adjustment of the threshold voltage for ONE/ZERO, or the setting of latching the input data at the rising or falling edge of the input clock. Usually, multiple bit inputs, such as 16/32/64/128/256 bits, are allowed.

Some come with the additional functionality of logic generation. As shown in Fig. 8.19b, output data is stored in memory in advance, and ONE/ZERO are sequentially output with the clock. Here, multiple bit outputs are also possible, and the output amplitude can also be selected. Some logic analyzers output data synchronously with an input clock from outside, whereas some others allow the user to select a frequency and output an internally generated clock.

Chapter 9
Measurement Techniques

We now know what measurement equipment to use. When we are ready to do measurements and actually step foot into the lab, we see a lot of parts that we have never seen before. What are we supposed to connect to what. . .?

9.1 Supply · Ground and the Return Path

9.1.1 Ground

In most cases with LSI, the voltage rather than the current is measured. Here, we always need to keep in mind "between where and where" that voltage is defined, and normally the voltage measured is in reference to "ground." However, even "ground" is a relative thing, and, for example, in Fig. 9.1, the power supply 1 is only ensuring the voltage of $V_1 - G_1$, and it does not guarantee the potential of V_1 with respect to the true ground. Therefore, the ground of supply 1 (G_1), the ground of supply 2 (G_2), and the ground of the oscilloscope must all be the same potential. To this end, the grounds of supply 1 and supply 2 are connected at another location. The closer these grounds are connected, the easier it is for them to become the same potential, and if possible, the grounds of supply 1 and supply 2 should also be shorted on the board. The potentials V_1 and V_2 relative to ground are more stable when G_1 and G_2 are connected internally within the chip. However, from the perspective of noise, the grounds cannot always be connected inside the chip, and the ground lines for the analog domain, digital domain, and IO are usually separated within the chip. Also, the same power strip should supply the power for the many measurement devices used when actually measuring the chip. The supplied voltage can differ from power strip to power strip, and it may become difficult to match the ground potentials for the used devices when the power is taken from separate strips.

T. Nakura, *Essential Knowledge for Transistor-Level LSI Circuit Design*,
DOI 10.1007/978-981-10-0424-7_9

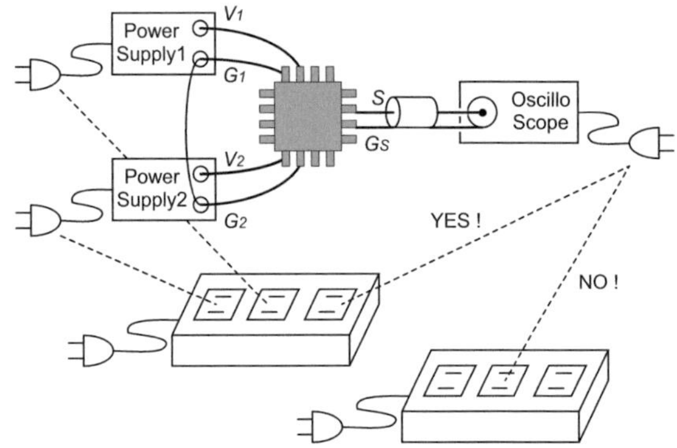

Fig. 9.1 Ground connections

9.1.2 *Supply and Decoupling Capacitance*

Variations in the supply voltage can cause malfunctions and therefore must be held to be as small as possible. The supply voltage varies due to the parasitic resistances and parasitic inductances on the supply lines. With the resistance, a supply voltage drop occurs due to IR, and not only does the long-term average voltage drop as $V_{DD} - I_{av}R$, but also an instantaneous undershoot occurs as $V_{DD} - I(t)R$. From the inductance, the long-term average voltage does not change from V_{DD}, but an overshoot or undershoot can occur instantaneously due to $L(di/dt)$. For synchronous digital circuit operation, delay errors (setup violations) due to instantaneous (on the order of one clock period) undershoot and delay errors (hold violations) due to instantaneous overshoot must be avoided. To achieve this, not only must the resistances and inductances be held small, but also the changes in the current flowing through the chip (di/dt) must also be suppressed. Adding capacitance is effective for this. As shown in Fig. 9.2a–c, by attaching capacitance, the instantaneously varying current is supplied by the capacitance, while the stationary average current required flows through the resistances and inductances, avoiding the effects of the parasitic inductors and therefore suppressing instantaneous supply voltage changes. This capacitance is called the decoupling capacitance.

It is desirable to place a large amount of decoupling capacitance as close to the circuit as possible. Adding on-chip capacitance to within the chip leads to an increase in the chip area, and therefore, off-chip capacitance is connected on the board, immediately next to the supply pin outside of the package.

As types of off-chip capacitance, there are electrolytic capacitors and ceramic capacitors. In general, electrolytic capacitors have larger capacitances. The impedance of a capacitor is $Z = 1/j\omega C$, and in an ideal capacitor as shown in Fig. 9.2d, connecting a capacitor with larger capacitance leads to a smaller impedance, better suppressing supply variations. However, in reality, both electrolytic and ceramic capacitors have parasitic inductances at the terminals as shown in Fig. 9.2e. At frequencies higher than the resonance frequency $f_{\text{res}} = 1/2\pi\sqrt{LC}$, the impedance rises and the supply variation suppression

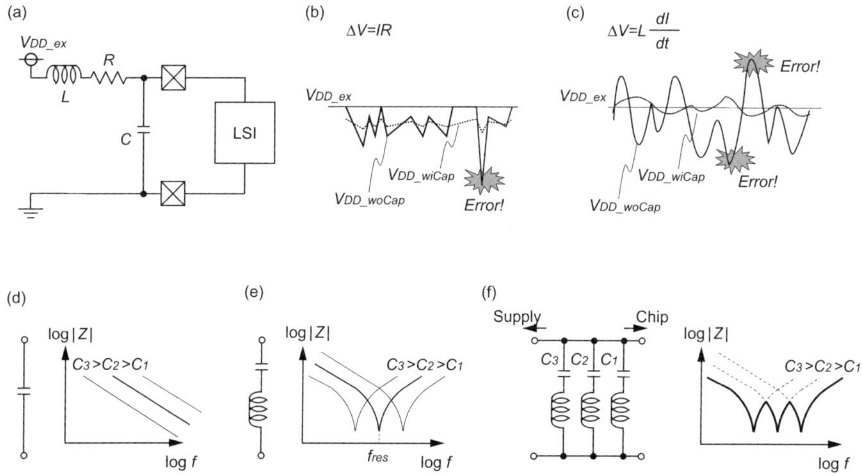

Fig. 9.2 Decoupling capacitance

effects diminish. To realize a low impedance in a wide frequency range, capacitors with various values are connected in parallel as shown in Fig. 9.2f. By absorbing low-frequency noise with larger capacitors and high-frequency noise with smaller capacitors, noise is absorbed across a wide frequency range. In this case, the smaller capacitors for the higher frequencies should be placed closer to the chip.

9.1.3 Return Path

Currents always flow in loops. When current or charge flows out of the positive terminal of a supply, the same amount of current or charge will flow into the negative terminal.

For example, we will consider the charge and discharge of the capacitance in an inverter. We often see the explanation that a charge of Q flows out from the supply during the charge phase as shown in Fig. 9.3a and that discharge through the NMOS during the discharge phase as shown in Fig. 9.3b. However, we can explain this from the loop current perspective in more detail as follows. During the charge phase, a charge of Q is supplied to the top plate of the capacitor by the supply as shown in Fig. 9.3c, and at the same time, a charge of $-Q$ accumulates on the bottom plate of the capacitor. That is, a charge of Q flows from the bottom plate to ground. The overall picture is that the charge appears to be flowing from the supply, through the capacitor, and into ground. Although direct currents do not flow through capacitors, the charge occurs with an alternating current which does flow through capacitors. The virtual current that runs through the capacitor here is called the displacement current (from your electromagnetics class, $\nabla \times \boldsymbol{B} = \mu_0 \boldsymbol{i} + \mu_0 \epsilon_0 \frac{\partial \boldsymbol{E}}{\partial t}$). During discharge, as shown in Fig. 9.3d, the charge that had accumulated on the top

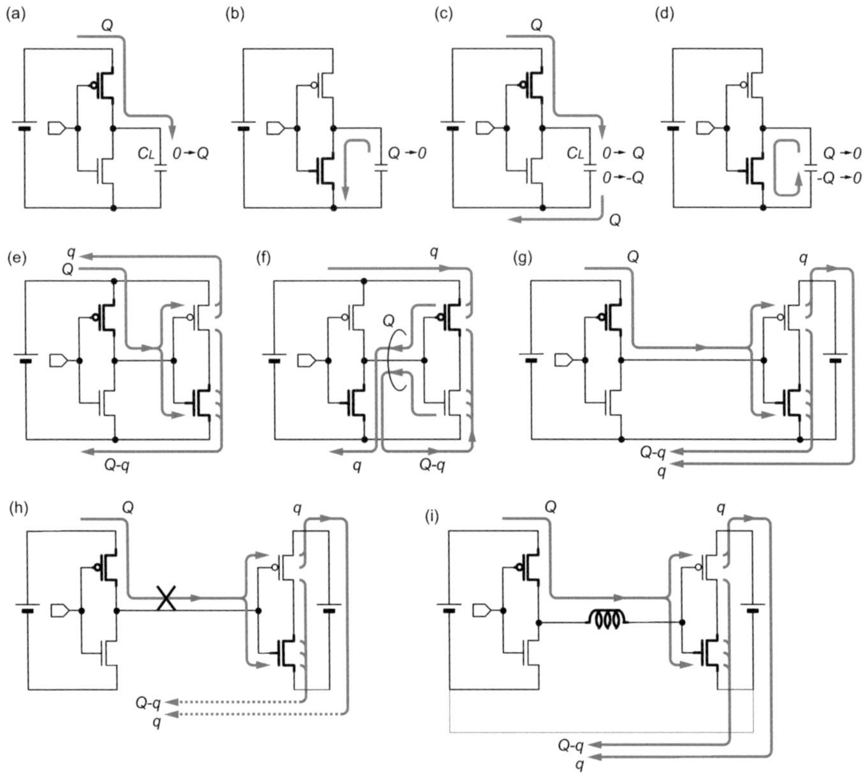

Fig. 9.3 Return path

and bottom plates of the capacitor neutralizes through the NMOS. Here, the charge flowing from the power supply is zero, and naturally, the charge flowing through the ground terminal is also zero (we are ignoring the charge necessary for the switching of MOS gates here).

The transfer of charge while an inverter is driving the next inverter stage, for the case when the output voltage of the first stage (the input voltage of the second stage) goes from LOW to HIGH and the output voltage of the second stage moves from HIGH to LOW, i.e., the PMOS of the first stage and the NMOS of the second stage are ON, is shown in Fig. 9.3e. Charge is injected into the G-S, G-B, and G-D capacitances of the second stage PMOS from the first stage PMOS, but the charge q on the G-S and G-B capacitances returns to the power supply, while the charge on the G-D capacitance flows to ground through the second stage NMOS that has turned ON. Charge is injected into the G-S, G-D, and G-B capacitances of the second stage NMOS by the first stage PMOS and flows to ground. Now, a charge of Q flows from the first stage PMOS, but a charge of q returns from the second stage PMOS. Therefore, the charge flowing out of the power supply is $Q - q$, and naturally it follows that the charge returning to ground is also $Q - q$. When the switching is reversed as shown in Fig. 9.3f, a charge of q flows out of the supply and q returns to ground.

Let us now compare the cases shown in Fig. 9.3c and e. The charge flowing through the first stage PMOS is Q in both cases, but the power consumption in Fig. 9.3c is QV while it is $(Q - q)V$ in Fig. 9.3e. What is actually going on here? If we compare the case of inverted input in Fig. 9.3d, f, in Fig. 9.3d the charge Q on C_L simply cancels out through the NMOS and the power consumption is zero, whereas in Fig. 9.3f a charge of q flows from the supply through the second stage PMOS and a power consumption of qV is generated. Therefore, the overall power consumption matches for a sequence of HIGH → LOW → HIGH transition cycles. In addition, as mentioned in Sect. 1.2.2, although we often see the second inverter in the circuit of that in Fig. 9.3e being depicted as a single capacitor to ground as that in Fig. 9.3c, this is a simplified description of the operation and differs from reality. We must also be wary of the fact that the transistor capacitances change depending on the bias voltages.

If a separate power supply is to be used in the first and second stages as in Fig. 9.3g, the charge q that is pushed out from the second stage PMOS flows into the supply (the charge provided by the other power supply decreases) and returns through the common ground. Therefore, Q flows through the signal line and Q returns through ground.

Now, when a current/charge flows from the positive terminal of a power supply to an LSI, the same amount of current/charge flows from the LSI into the negative terminal of the supply. Similarly, a current/charge that flows through a signal line also returns through ground. This route that the current takes is called the return path. When the grounds are not connected as shown in Fig. 9.3h, no current flows from the first stage inverter to the second stage because there is no return path. When the second power supply is connected to ground at some faraway location as in Fig. 9.3i, although a return path exists and a current will flow in a loop, the larger the loop area, the more difficult it is for the current to flow. In fact, it will appear as if a large inductance proportional to the loop area has been inserted in the signal line and the high-frequency characteristics will degrade. To suppress the inductance and allow the current to flow more easily, the area of the loop must be minimized. Therefore, the supply/ground lines or signal/ground lines should be wired as close as possible. Be careful not to disconnect the return path especially when the signal routing becomes very long.

9.2 Various Components

9.2.1 Connectors and Cables

There are various types of cables and connectors for signal transmission, and the correct ones must be chosen depending on the transmitted and received signal frequencies. Connectors that are often used (the author regularly uses) are shown in Fig. 9.4. By understanding the characteristics of these connectors and cables, they can be used properly for conducting measurements.

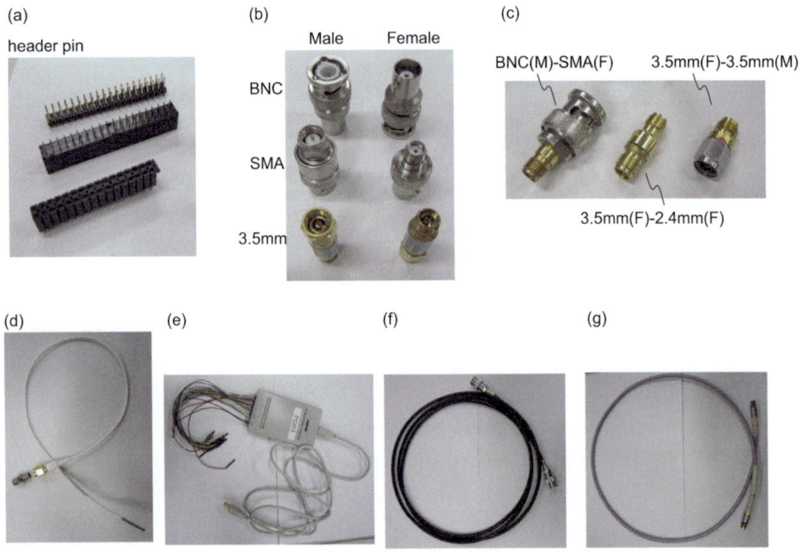

Fig. 9.4 Various connectors and cables

9.2.1.1 Header Pins

The objects in Fig. 9.4a are called header pins and are mostly used for slow (~100 MHz) digital signal transmissions. They are often used to transmit multiple digital signals, for example, 16 or 64 bits. The pin separation distance is standardized to be 2.54 mm (a unit of distance called mil is where this number comes from, where 1 [mil] = 0.0254 [mm]). I really hope they stop using these miles, yards, inches, and other non-MKS units). The object in Fig. 9.4e is a device called the "pocket generator" that receives commands from a computer via USB and outputs 16 bit signals from the header pins. Even in this case, it is important not to forget to connect the ground line as the return path as well as to match the voltage levels of the transmitting and receiving ends. In this example, a single ground line is coming out for 16 bit signal lines.

9.2.1.2 BNC, SMA, 3.5 mm, 2.4 mm, 1.85 mm

The connectors in Fig. 9.4b are for analog and high-speed signal transmission, and there exist standards such as BNC, SMA, 3.5 mm, as well as 2.4 and 1.85 mm which are not shown. The frequency ranges of use are, as a rule of thumb, header pins, 100 MHz; BNC, 2 GHz; SMA, 22 GHz; 3.5 mm, 26 GHz; 2.4 mm, 50 GHz; and 1.85 mm, 67 GHz. Figure 9.4f, g shows BNC and SMA cables. Their characteristic impedances are 50 Ω. In most cases both ends of the cable are male connectors, and the device side has female connectors. The center of these connectors is the signal lines and the outer parts are ground, so that the ground potential is matched and the return path is ensured.

9.2.1.3 Conversion Connectors

Figure 9.4c shows a conversion connector. Due to various reasons such as the equipment being used and the available cables and accessories (discussed later), converting connectors can become necessary. SMA-3.5 mm, 2.4 mm–1.85 mm can be connected without conversion, and some even say that SMA-3.5 mm connections even have better characteristics than SMA-SMA. The space between signal and ground in an SMA is filled with a dielectric, whereas the 3.5 mm is insulated by air.

9.2.1.4 Cables

A conversion connector for a header pin and SMA is shown in Fig. 9.4d. A line for signal and a line for ground are running in parallel, and two female connectors are separated by 2.54 mm on the header pin side. The SMA side is male in this example, so a female-female conversion connector is often used to connect to the signal-generating equipment through an SMA cable. Fast signals cannot be transmitted because of the header pins. A lot of these connectors may be necessary at once, and sometimes the bottleneck in measurements can be how many of these are available.

Figure 9.4e shows a USB-type pulse generator. Header pins are used to output multiple low-speed digital signals.

Figure 9.4f shows a BNC-type transmission cable, with a characteristic impedance of 50 Ω. Generally, the cable terminal is male, and the equipment terminal is female.

Figure 9.4g shows an SMA-type transmission cable, with a characteristic impedance of 50 Ω. Generally, the cable terminal is male, and the equipment terminal is female.

9.2.2 *Accessories*

In measurements, we wish to operate on signals by attenuation, cutting the DC components, adding DC components, or dividing signals into two. For this, special accessories such as those shown in Fig. 9.5a are used. Normally, the characteristic impedances are $Z_0 = 50\,\Omega$.

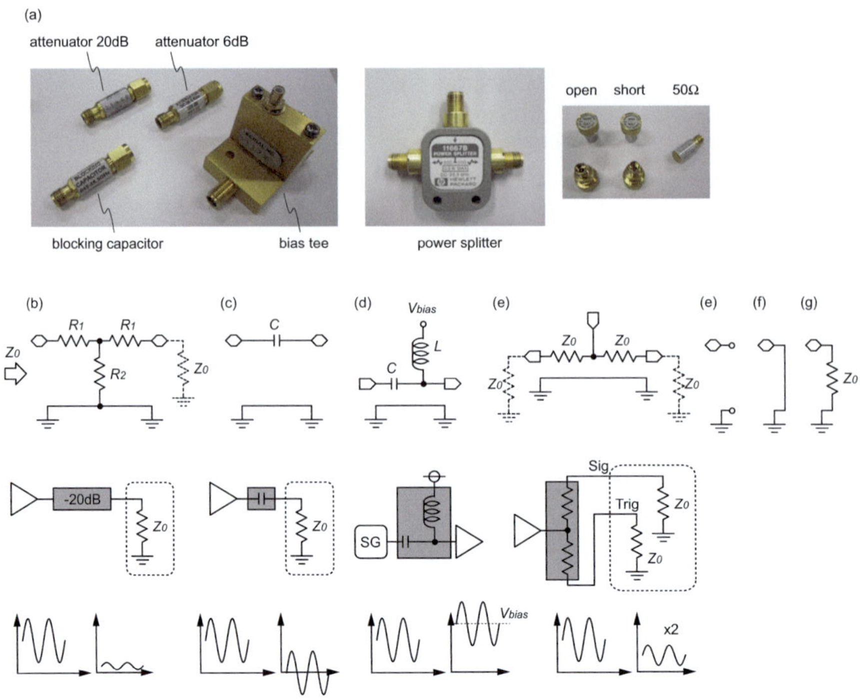

Fig. 9.5 Accessories

9.2.2.1 Attenuators

Figure 9.5b shows an equivalent circuit of an attenuator. The signal is attenuated while the impedance is matched. This is used when, for example, the oscilloscope input voltage range is 0.5 V, so the signal with 2 V amplitude is measured after being attenuated by 1/10. The impedance looking from the left is Z_0. In addition, if the voltage V_{OUT} on the load Z_0 is α (<1) times the input voltage V_{IN}:

$$Z_0 = R_1 + (R_1 + Z_0)//R_2 \tag{9.1}$$

$$\frac{V_{OUT}}{V_{IN}} = \frac{R_2//(R_0 + Z0)}{R_1 + R_2//(R_1 + Z_0)} \cdot \frac{Z_0}{R_1 + Z_0} = \alpha \tag{9.2}$$

which means:

$$R_1 = \frac{1-\alpha}{1+\alpha} Z_0 \tag{9.3}$$

$$R_2 = \frac{2\alpha}{1-\alpha^2} Z_0 \tag{9.4}$$

For $-6\,\mathrm{dB} \rightarrow \alpha = 1/2$, and $-20\,\mathrm{dB} \rightarrow \alpha = 1/10$.

9.2.2.2 Blocking Capacitors

Figure 9.5c shows an equivalent circuit of a blocking capacitor, which cuts out the DC contents of the signal and only allows the AC portion to pass through. A large capacitor will seem like a short to the fast AC portion ($Z_C = 1/j\omega C$). This is used when, for example, the output of an amplifier is observed by an oscilloscope terminated to G_{ND} by 50 Ω without changing the output bias point. Ideally, the larger the capacitance, the lower the transmittable signal frequencies are. However, large capacitive elements have poor high-frequency characteristics, and if a bandwidth at a high frequency of 3.5 mm (~20 GHz) is to be transmittable, the lower bound on transmittable frequency turns out to be around 50 MHz.

9.2.2.3 Bias Tees

Figure 9.5d shows an equivalent circuit of a bias tee. A DC component is added to the AC component of the input signal and then sent as output. This is used, for example, when a bias of $V_{DD}/2$ is added to the output of a signal generator. The AC component of the input signal passes through C and the DC portion is cut out. As long as L is large enough, the AC portion that has made it through will be transmitted to the output without being affected by the L, and the DC voltage added externally will become the offset.

9.2.2.4 Power Splitters

Figure 9.5e shows an equivalent circuit of a power splitter. A signal is split into two while impedance is matched. If each output terminal is terminated with Z_0, the impedance looking in from the input will be $2Z_0//2Z_0 = Z_0$, and therefore the impedance is matched and no reflections will occur. However, the output voltage will be halved. This is used, for example, when the CLK signal output is observed with a sampling oscilloscope and one is used as the signal while the other is used as the trigger.

9.2.2.5 Termination

That shown in Fig. 9.5e–g is used to terminate the end of a cable with an open, a short, and 50 Ω, respectively. The 50 Ω termination is used for the calibration of network analyzers, as well as when, for example, a circuit has three output terminals but the oscilloscope only has two input terminals and the characteristics would change if the third terminal is left open.

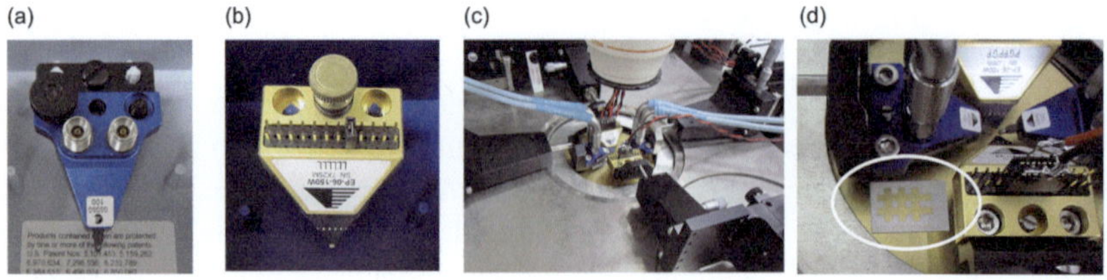

Fig. 9.6 Probers

9.2.3 *Probes*

If a chip is packaged with QFP, fast signals in the GHz range cannot be input or output due to the parasitic inductances of the bonding wires and lead frames. At the laboratory level, measurements are sometimes made by directly placing needles on the wafer itself, as shown in Fig. 9.6.

9.2.3.1 Probers for High-Speed Signals

Figure 9.6a shows a wafer prober for high-speed signals. It has a 3.5 mm connector and is manufactured with a characteristic impedance of 50 Ω to the end of the needle. There are two signals in this picture, each shielded by G_{ND}, resulting in a total of five needles in the order of GSGSG. The pad arrangement on the chip side must be in this order as well. Also, because a 50 Ω termination resistance cannot be connected to the end of the needle, an input buffer with a built-in 50 Ω termination resistance must be used.

9.2.3.2 Probers for Low-Speed Signals and Power Supplies

Figure 9.6b shows a prober for low-speed signals and power supplies. Signals, power, and ground are given to the end of the needle through the header pin. There are some that have a capacitance of a few pF between the two terminals in the needle, and these pins are to be used for supplying power and ground.

9.2.3.3 Pad Pitch and Pin Count and Items for Adjusting Needles

The interval between needles in a probe can be specified when making (ordering) probes. Of course, the number of needles can be specified as well. For circuits that are measured by probing, it is often the case that the transistor count is small and thus the chip area is mostly taken up by the pads and the core layout is rather empty. With the advancement of processes and the increase in the cost per chip area, we would

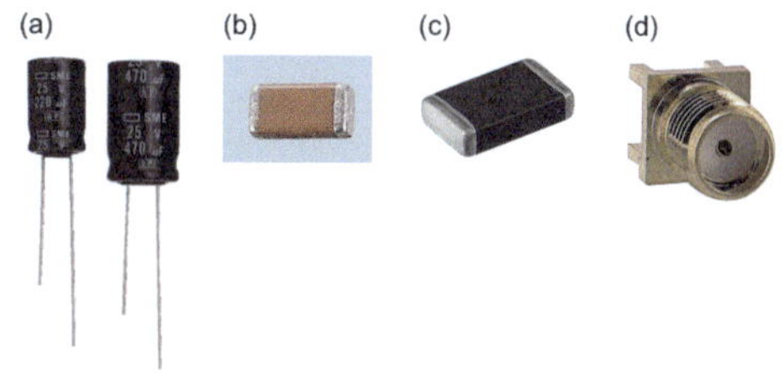

Fig. 9.7 Assembly components

like to make the pads as small as possible and the pad interval as small as possible to minimize the area. In the past, pad pitches (needle-to-needle distances) were 150 or 100 μm, but in recent years, there are 60 μm pitches as well. To measure, needles are placed in contact with pads while looking under a microscope as shown in Fig. 9.6c, and skillfulness of the hands is required. This procedure of contacting the needle is actually fairly difficult with a 60 μm pad pitch. When the probe is not parallel with the pads, one needle could be in contact with the pad while the other is not. An adjustment item for contacting needles as shown in Fig. 9.6d is used to flatten the probe first before actually placing the needles on the pads.

9.2.4 Assembling Components

When packaging the chip and assembling the board, various components such as those shown in Fig. 9.7 are mounted, and signal inputs and outputs as well as supply voltage stability are planned and attempted. Below, some components the author often uses are presented, but there is a huge variety of components, so it is suggested that you hunt around in Akihabara or look for the components you need on appropriate websites.

9.2.4.1 Electrolytic Capacitors

An electrolytic capacitor is shown in Fig. 9.7a. These have capacitances of several μF to several mF. High-frequency characteristics are poor due to parasitic impedance effects. The terminals have a positive and negative side, and the longer leg is connected to the plus side. If these are switched, the capacitor will blow with a loud "pop" (I've done this before).

9.2.4.2 Chip Capacitors

A chip capacitor is shown in Fig. 9.7b. These have capacitances of several pF to several hundred nF. There are several standardized sizes, and while smaller ones will lead to smaller PCBs, they are more difficult to handle. For use in the lab, ones with a long side of about 1.5 mm should be used, because components smaller than that will cause trouble when handling with tweezers. Parasitic impedances are low, and frequency characteristics are better than electrolytic capacitors. Even then, they have frequency characteristics as depicted in Fig. 9.2. Their unique labeling indicates the capacitance values. For example, labels of "101" or "104" indicate 10×10^1 [pF] = 100 [pF] and 10×10^4 [pF] = 100 [nF], respectively.

9.2.4.3 Chip Resistors

A chip resistor is shown in Fig. 9.7c. These are used as termination for high-speed input buffer pins without any internal termination resistance. If the resistance label is "501," this means that the value is 50×10^1 [Ω].

9.2.4.4 Surface Mount Type SMA Connectors

A surface mount type SMA connector, for supplying high-speed signals to the board, is shown in Fig. 9.7d. Gold-plated ones are easier to solder with.

9.2.5 Shield Room

Various electromagnetic waves are flying around in the air, with cellular phones and wireless LAN being the representative ones. As a result, some noise can occur in a signal within a chip. As a designer, it is important that the circuits operate properly even when these radio waves come flying. However, there are cases in delicate measurement when we would like to cut out all the noise from the outside world to observe the effects of the circuit internals and the circuit externals separately. A shield room is a space surrounded by metal and cuts off any electromagnetic waves coming from the outside. There are some large enough for people to enter, as shown in Fig. 9.8a, and some that are only large enough for equipment as in Fig. 9.8b. If these are not available, a cardboard box wrapped in regular aluminum foil will suffice. You must be careful to make sure that the aluminum itself is not conducting and shorting unintended signals. In either case, a simple way to check the shield is to place your cell phone inside and call it using a friend's phone. If the phone cannot be reached, the shield is OK. By the way, in the author's experience, the cell phone was disconnected in the cases shown in Fig. 9.8a and the aluminum foil box, but the phone actually did ring inside the shield box shown in Fig. 9.8b . . .

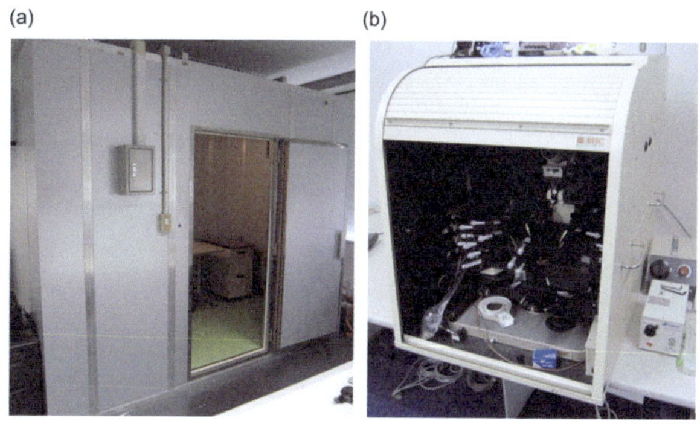

Fig. 9.8 Shield rooms

9.3 Assembly Examples

Several boards that the author has built for measurements are shown in Fig. 9.9. All of them are achievements gained through battles using tweezers and soldering irons. They come out beautifully if in the beginning solder is used abundantly and later the excess is sucked up using desoldering lines.

Figure 9.9a shows the board for when substrate noise was measured. The required pattern is drilled into a glass-epoxy board with several dozen μm of copper film on top with a drawing drill, and the chip is directly soldered onto the copper plate. Most of the copper plate surface is utilized as ground. The supplies Vdd_** are given through lead lines from the voltage supply equipment to the board in pairs of supply and ground. An "island" for the supply is created on the supply pin side and the supply line is soldered to this island, while ground lines are soldered to the outside of this island. To remove noise across a wide frequency range, several types of chip capacitors are connected in parallel between the chip and the outside seas (refer to Fig. 9.2f). High-speed digital signals and high-speed analog signals are extracted using 50 Ω SMA connectors, which are connected to oscilloscopes with input impedances of 50 Ω. The output buffers are designed with this in mind. Here, a thin transmission line with characteristic impedance of 50 Ω is used between the SMA connector and the pin. The inner core and surface are separated right next to the package pin, and the core is connected to the pin while the surface is soldered to the board ground. Also, because the measurements will be taken with a sampling oscilloscope, a CLK/32 signal for triggering is internally generated.

Figure 9.9b shows a similar case to that in Fig. 9.9a, but because the *allORHalf* signal output was a slow digital signal of several MHz, it was observed by hooking the prober of the oscilloscope onto a wire connected to the pin. A CMOS inverter for low-speed signals is used for the output buffer, and the oscilloscope input impedance is set to 1 MΩ instead of 50 Ω.

In Fig. 9.9c, surface mount type SMA connectors are utilized to input and output digital signals of several hundred MHz. Because these signals were not sensitive,

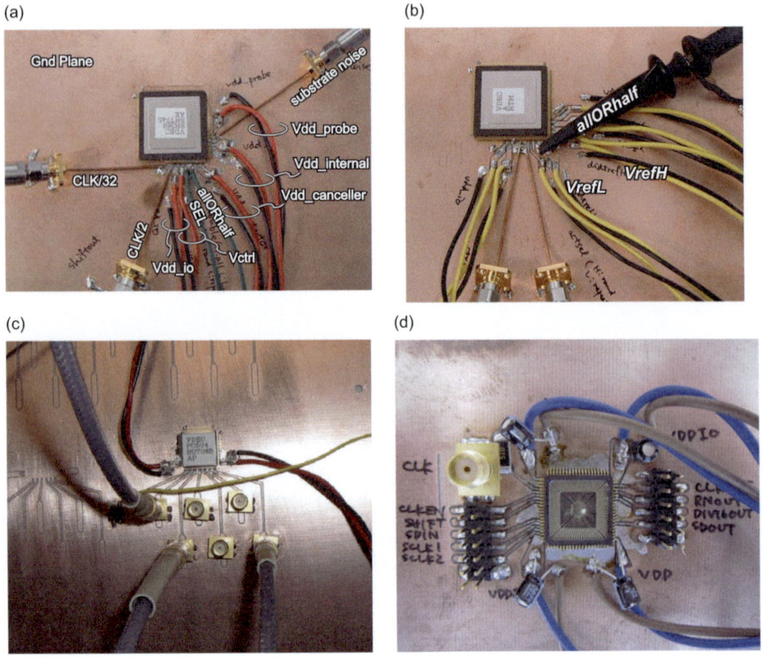

Fig. 9.9 Assembly examples

the patterns from the SMA connectors to the chip are etched without worrying too much about the characteristic impedances. A termination resistance of 50 Ω is built in to the input buffers, and the CLK signal is input with a signal generator and a bias tee to add a bias voltage of $V_{DD}/2$.

In Fig. 9.9d, we needed to input and output a large number of low-speed digital signals from the logic analyzer. Thus, header pins are used to input signals into input buffers without 50 Ω termination, and the outputs from the chip are also connected to the logic analyzer input through header pins. As for the external CLK, a 50 Ω SMA is used and is input from a signal generator through a bias tee to an input buffer with 50 Ω termination.

In Fig. 9.6c, d, measurements are made using probes. GSGSG high-speed probes are used for high-speed signaling, and supply voltages are supplied with low-speed probes. At the bottom right of Fig. 9.6d, "grandchild" header pins for extension purposes are placed on top of the probe header pins, and decoupling capacitances with electrolytic capacitors are soldered between supply and ground.

As a common point of caution, supply lines should always be supplied in pairs with ground, and signal lines should also be connected in pairs with ground. This ensures both a return path and a common ground potential.

9.4 GPIB, Measurement Automation, and C Programming

When making measurements on a chip, the supply and measurement equipment such as the signal generators are first connected with the chip, and waveforms are observed with, for example, an oscilloscope. In most cases, various operation modes such as the voltages and CLK frequencies are swept while the changes in waveforms or spectra are measured. For example, we consider the case of measuring the dependence on the supply voltage of a ring oscillator that can switch between five-stage and seven-stage modes. First, you set the supply voltage to 0.5 V and look at the waveform, then the markers are adjusted to make readings. You make note of the frequency in your notebook, and repeat the experiment for 0.6, 0.7 V, ... all the way up to 2.5 V (20 times!). You conduct this experiment for both the five-stage and seven-stage modes. You leave the lab to enter your data into an excel sheet to draw a graph, but something doesn't look right, because you've made a mistake. You also notice that the results change depending on whether you made the measurement in the morning or at night. You would like to go through the measurement process again, but it's so tedious...

Automation should always be considered when making measurements. Most measurement equipment will have GPIB ports in the back, and so they can be controlled with C programs from a PC through a USB-GPIB conversion cable. More recent equipment can even have USB ports instead of GPIB ports. Through these communication ports, the voltage of a source can be set and the current value can be read out, or the waveform of an oscilloscope can be exported as numerical data of the time and voltage. A program can even be written to extract the oscillating frequency from this data. Measurement should be conducted with the mind-set of "measurement $\approx$ programming," and about 70 % of your time in the lab should be spent in front of a computer display writing programs for control.

Figure 9.10a shows a conversion cable and Fig. 9.10b, c the front and back of a voltage supply, which has a GPIB port in the back. Figure 9.10d shows an actual example of a measurement. This is measuring the chip in Fig. 9.9d, and the supply, logic analyzer, oscilloscope, and signal generator are all connected through GPIB. A C++ program was written to make the following measurements while sweeping several supply voltages and frequencies: feed in CLK from the signal generator, input a specified signal from the logic analyzer and measure the output signals, send the results to the PC, send the oscilloscope waveforms also to the PC, analyze the measurement results, and output the corresponding waveforms from the logic analyzer.

If the measurement equipment was purchased recently, a CD with a C++ library to control the device with a C++ program should be included. A list of commands to send to the equipment can be found by referencing the Programmer's Guide in the equipment manual. You should study C++ programming with an appropriate textbook. However, C++ isn't always necessary, and sometimes C will suffice to write control programs. A program to set the supply voltage and read the current value at that time is listed below. The code is somewhat long because classes, a

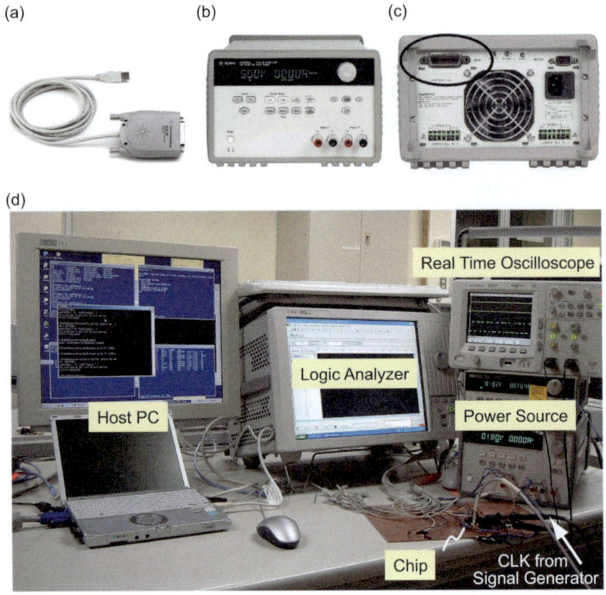

Fig. 9.10 GPIB control

feature of C++, are used, but a few lines of code are enough to set the voltage and read the current.

```
#include "sicl.h"
using namespace std;

#define VSOURCE_ADDRESS "gpib0,1"
const int VIOPORTNO = 1;

// -------------- class definition --------------
class Device_t
{
  public:
    Device_t(char* address);
    virtual ~Device_t(void);

  protected:
    INST _DeviceId(void);

  private:
    INST _deviceId;
};

class Vsource_t : public Device_t
{
  public:
    Vsource_t(char* address);
```

```
    virtual ~Vsource_t(void);

    void SetVolt(int channelNo, double volt);
    double GetVolt(int channelNo);
    double GetCurrent(int channelNo);

  private:
};

// -------------- method for Device_t --------------
Device_t::Device_t(char* address)
{
  _deviceId = iopen(address);
  if (_deviceId == 0)
    {
      cout << "The device " << address << " is not found."
           << endl;
      exit(1);
    }
}

Device_t::~Device_t(void)
{
  iclose(_deviceId);
}

// -------------- method for Vsource_t --------------
void Vsource_t::SetVolt(int channelNo, double volt)
{
  char command[MAXLINELENGTH];
  sprintf(command, "INST:SEL OUT%d\n", channelNo);
  iprintf(_DeviceId(), command);
  iprintf(_DeviceId(), "VOLTage %lf\n", volt);
}

double Vsource_t::GetCurrent(int channelNo)
{
  char command[MAXLINELENGTH];
  sprintf(command, "INST:SEL OUT%d\n", channelNo);
  iprintf(_DeviceId(), command);

  double current;
  ipromptf(_DeviceId(), "MEASure:CURRent?\n", "%lf", &current);

  return current;
}

// -------------- main --------------
int main(int argc, char* argv[])
{
  Vsource_t vsource(VSOURCE_ADDRESS);

  vsource.SetVolt(VIOPORTNO, 1.2);
  double current = vsource.GetCurrent(VIOPORTNO);
```

```
  cout << "current= " << current << endl;

  return 0;
}
```

Chapter 10
The Overall Design Procedure

Now that you've read this far along, you should have obtained the necessary knowledge for each phase of design. Here, the overall design procedure is summarized.

10.1 Before Starting Your Design

10.1.1 What Are You Making and Why

The points to take into consideration are completely different depending on whether the design is a large-scale CPU for super computers or it is a five-stage ring oscillator just to get used to the CAD tools. Priorities will again differ, based on whether the design is for a mass-produced (a million chips per month) final product or it is targeted more for publications in academia.

As a novice, you might be told to (forced to?) design circuits specified by your teachers or bosses. As you become more accustomed to design, you'll be able to suggest, "please let me design xx," or you might even be given unreasonable mandates such as "just make anything that will meet next month's shuttle." Once the design flow including how to use CAD tools has been mastered, in the end, the phase requiring the most wisdom and knowledge is thinking about "what to make." When you begin to feel this way, you can call yourself a true expert.

Equally important is to grasp "for what purpose" you will be doing the design. Here I don't mean "for world peace," or "to use in self-driving cars," but rather classifications as listed below:

1. To create a final product
2. For an engineering sample product before finalization
3. To determine whether or not to pursue product development
4. To obtain top data and present at an academic conference
5. To implement a new functionality and obtain patents or to present at a conference

T. Nakura, *Essential Knowledge for Transistor-Level LSI Circuit Design*,
DOI 10.1007/978-981-10-0424-7_10

6. For training the newcomer
7. To understand the circuit operations (only design is conducted, and no chip is fabricated)

In items "1. To create a final product" and "2. For an engineering sample product before finalization" above, measurements become large in scale, and various detailed specifications are determined. There are various test methods such as the ATPG, at speed test, scan test, 3σ statistical analysis, etc., which all have enough material to write several textbooks by themselves, and the specifications must be determined by consulting an expert in testing. In this text, we will limit our discussion to items 3, 4, and 5 where the story basically ends within the laboratory.

10.1.2 Determining the Final Image

10.1.2.1 Specifications

1. What is the supply voltage?
2. What is the operating frequency?
3. How much is the power dissipation?
4. What are the analog components?
5. How much PVT variation will be taken into consideration? Will you gamble on the resulting outcomes?

If the chip is low power consumption, then the design should take into consideration the reduction of the supply voltage to minimize the power consumption. Or, if the chip is high performance, the supply voltage would be determined by the process standard, and the power consumption would not be as much of a concern. As analog points of consideration, A/D and D/A require INL and DNL to be excellent, and PLLs would place the jitter characteristics under the most scrutiny. In these ways, each circuit has its own set of specific characteristics to consider. If the objective is "top data" or "implementation of new functionality," then the operating frequency is planned to be in the GHz range, but with PVT variation, there are cases where we will find out the exact results after fabricating the chip and making measurements.

10.1.2.2 Utilized Process and Design Deadlines

1. Determine the process technology based on performance and cost
2. One that has the closest deadline for which design is possible
3. Based on the shuttle schedule
4. Based on conference or dissertation deadlines
5. Sudden, unforeseen opportunities
6. CAD setting files

The circuit specs, process technology, and design deadlines are strongly correlated with each other, so the above points should be taken into consideration to find the optimal solution for your case.

10.1.2.3 Measurement Environment

1. Will we package the chip to measure or measure with wafer probing?
2. If we are going to package the chip, what do we do about the board? Can we reuse an existing one, or will we have to manufacture a new one?
3. What equipment will we use to make measurements? Oscilloscopes? Spectrum analyzers?

Depending on the target specifications of the circuit, you can potentially run into the problem of not owning equipment that has such high-speed capabilities or not owning any equipment that can make the type of required measurements at all. If this is the case, equipment must be purchased or rented, or, otherwise, measurement circuit compatible with the available equipment must be thought up, designed, and embedded into the circuitry. If a new board is being fabricated, the board design schedule must also be taken into consideration.

10.1.3 Determining CAD Tools

Once the general framework of the design has been determined, the CAD tool to be used for the design must be chosen. Various companies sell their own CAD tools, even though they may have similar functionality. It is best to use tools that you are accustomed to, but whether or not you can use a particular tool depends on whether the appropriate setting files are provided. If the setting files are only provided for tools that you have never used, then your only option is to get used to the new tool. Starting off with a new tool can be simple, but troubleshooting error messages will take some time with trial and error.

Various setting files necessary for CAD types are listed below. These are representative examples, and depending on the tool, these setting files can be bundled into a single file or can come separately. Also, there are some that you can create on your own, and the provided files are customized and used as necessary. However, when customizing, it is necessary to ensure consistency within the group and that no conflicts arise in the final data. Also, when data is sent back and forth between tools, attention is required in the details of data consistency, as is represented in the exemplary problem of case sensitivity. Utmost care is especially necessary when the final GDS data is merged and cell names conflict, in which case the cell names must be changed. In some cases, you might have to write your own scripts.

For example, in my case, a SPICE netlist is generated from a circuit editor, a script that I have written inserts AD/AS/PD/PS into the SPICE netlist, and then

SPICE simulations are run. Also, I have created a command that allows me to specify a cell name, output a GDS file, and run LVS and DRC, which allows me to run LVS/DRC easily from the command line.

Schematic Editor	• Basic libraries, such as transistor symbols • File for display color settings • File for shortcut settings • Setting file for calling the simulator from the schematic editor • Setting file for cross-referencing with the layout editor
SPICE Simulator	• SPICE parameter files • Options for cross-probing SPICE simulation results from the schematic editor
Layout Editor	• Setting file for display colors and patterns • File for mapping the mask layers (GDSII layers) to editor layers • File for shortcut settings • Setting file for cross-referencing with the schematic editor • Setting file for displaying LVS/DRC errors
LVS/DRC/ERC/Antenna	• LVS rule files • DRC rule files • ERC rule files • Antenna rule files • Dummy metal generation rule files and density check rule files
Interconnect RC Extraction	• LVS rule files • Setting files for the cross-sectional structure of interconnect or reference tables • Option setting files

10.2 Checking Transistor Characteristics

After the process technology is determined and the set of rule files is obtained, the first thing to do is to check the transistor characteristics.

10.2.1 SPICE Parameters

First things first: the SPICE parameters. Let us go find the SPICE parameter file and take a look inside.

```
.MODEL NLP NMOS
+ LEVEL = 53
+ VERSION = 3.2
+ TOX = 10e-9
+ ....

.MODEL NHP NMOS
+ LEVEL = 53
+ VERSION = 3.2
+ TOX = 9e-9
+ ....
```

In the above example, two types of model names for NMOS (`NLP` and `NHP`) are available for use, and judging from the names, they must be low power and high performance. We can, for example, decide to use the high performance `NHP` rather than the low power because we are placing greater significance on the operating frequency in this design. Additionally, we see from `LEVEL = 53` that this is using the BSIM3 model, and we note that **HDIF** can be used, but gate resistance and gate leakage are not taken into account. The correspondence between LEVEL and the model, as well as the meanings of parameter values, is described in a special manual, which should have been copied when the tool was installed. For example, in the case of HSPICE, a detailed description can be found at `$INST_DIR/hspice/docs_help/hspice_mosmod.pdf`.

Also, you should be able to find library descriptions `.LIB` within the SPICE parameter file or in another separate file located nearby.

```
.LIB NT
.PARAM tox = 10e-9
.ENDL

.LIB NS
.PARAM tox = 11e-9
.ENDL

.LIB NF
.PARAM tox =  9e-9
.ENDL
```

In this example, there are three types `NT, NS, NF` of libraries for variation for the NMOS, and judging from the names, they must be typical, slow, and fast. We can decide to use `NT` for design and try `NS, NF` as necessary.

10.2.2 DC Characteristics and Inverter Delay

The $I_D - V_D$ curves can be drawn for both the NMOS and PMOS. The simulations should be run with minimum L, $W = 1\ \mu$m, V_D from 0 to V_{DD} in 0.01 V steps and V_G from 0 to V_{DD} in 0.1 V steps. The graph in Fig. 10.1 should be posted in a visible location. During the design phase, this graph should be referenced often. Also, if

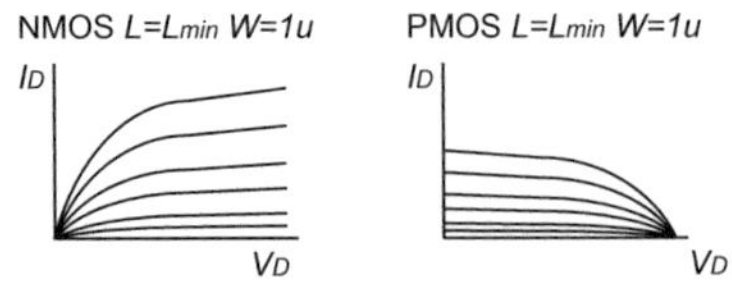

Fig. 10.1 DC characteristics

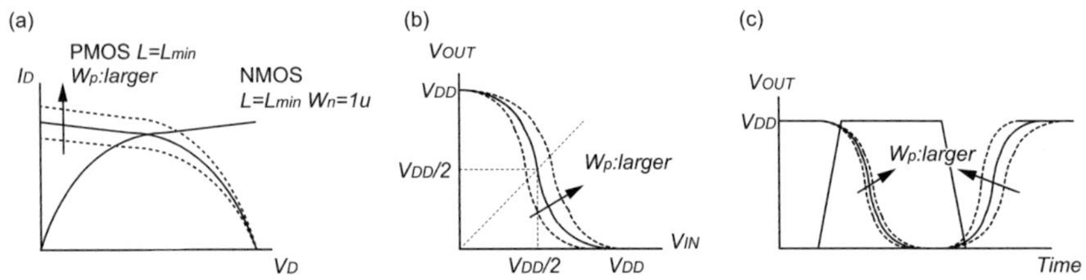

Fig. 10.2 Determining W_p

this graph is generated in the same format all the time, then in the future when designing with different process technologies, it is convenient to be able to compare the approximate performance of each process.

```
.OPTION POST=2 POST_VERSION=2001
.param mvdd = 1.8

.DC VD 0 mvdd 0.01 VG 0 mvdd 0.1

VD  D  0 DC mvdd
VG  G  0 DC mvdd
VS  S  0 DC 0
VB  B  0 DC 0
m1 d g s b NHP L=0.18u W=1u

.include "../rules/vdec1.par"
.lib "../rules/vdec1.lib" NT
.lib "../rules/vdec1.lib" PT

.end
```

Next, transistor sizing for basic gates is determined. Of course, various transistor sizes will be used in various situations, but we first design the most basic sized inverter. The NMOS width W_n should be in the ballpark of 5–10 times $L_{\min}$. It will be smaller for targeting low power dissipation and larger for high operating frequency. In a 0.18 μm process, a fast design would be around 2 μm to have a nice round number. The PMOS width W_p is determined by balancing with the NMOS. The NMOS and PMOS drain currents can be matched (Fig. 10.2a), the ideal inverter threshold can be tuned to $V_{DD}/2$ (Fig. 10.2b), or the L→H and H→L delays ($V_{DD}/2 \rightarrow V_{DD}/2$) can be matched (Fig. 10.2c). There is no one W_p that satisfies all of these conditions, but (a)–(c) are all considered and a round number should be chosen. While you are at it, the inverter delay should also be checked.

Furthermore, the approximate capacitance values should be looked at. The values for `TOX, CJ, CJSW, CJGATE, PB, PBSW, PHP, MJ, MJSW, MJGATE`, and `HDIF` should be found from the SPICE parameter file and

$$C_{ox} = \epsilon_r \epsilon_0 L_{\min} \times 10^{-6} / \mathbf{TOX} \tag{10.1}$$

$$C_{j0} = \mathbf{CJ} \times \mathbf{HDIF} \times 2 \times 10^{-6} \tag{10.2}$$

$$C_{jsw0} = \mathbf{CJSW} \times 10^{-6} \tag{10.3}$$

$$C_{j\mathrm{gate}0} = \mathbf{CJGATE} \times 10^{-6} \tag{10.4}$$

$$C_{\mathrm{total}0} = C_{ox} + C_{j0} + C_{jsw0} + C_{j\mathrm{gate}0} \tag{10.5}$$

as well as

$$C_{jV_{DD}} = C_{j0} \times (1 + V_{DD}/\mathbf{PB})^{-\mathbf{MJ}} \tag{10.6}$$

$$C_{jswV_{DD}} = C_{jsw0} \times (1 + V_{DD}/\mathbf{PBSW})^{-\mathbf{MJSW}} \tag{10.7}$$

$$C_{j\mathrm{gate}V_{DD}} = C_{j\mathrm{gate}0} \times (1 + V_{DD}/\mathbf{PHP})^{-\mathbf{MJGATE}} \tag{10.8}$$

$$C_{\mathrm{total}V_{DD}} = C_{ox} + C_{jV_{DD}} + C_{jswV_{DD}} + C_{j\mathrm{gate}V_{DD}} \tag{10.9}$$

should be calculated to have an idea of the rough capacitance per $W = 1\,\mu$m at reverse bias and zero bias (refer to Fig. 1.9). `CJ` is expressed in [F/m^2] and `CJSW` and `CJGATE` in [F/m]. Also, if `CJGATE` is not defined, it is treated as equal to `CJSW`.

10.3 Checking the General Flow

10.3.1 Schematic Editor

While looking at the appropriate manuals, you can draw the schematic for an inverter, with input and output pins. Make sure to output the supply and ground as terminals as well.

Then, you should do as follows: generate the input file for SPICE simulations, output the netlist from the schematic editor, run the SPICE simulator, and apply settings so that simulation waveforms can be cross-probed from the schematic, as shown in Fig. 10.3. Next, create a symbol for the inverter, design a circuit with five stages of the symbol, and simulate to observe the waveforms. This is all to make sure that hierarchical design can be done, as well as to simulate the inverter delay value.

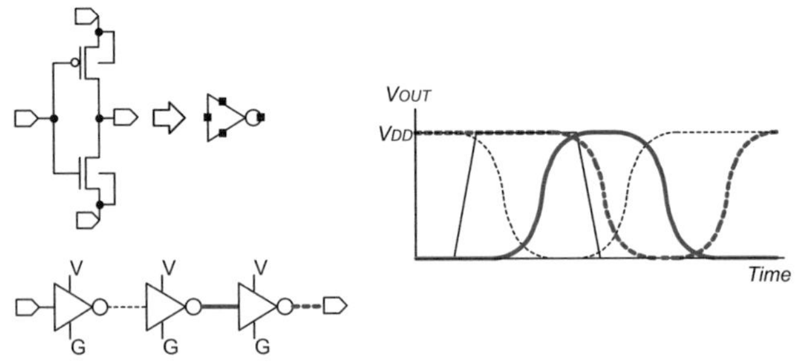

Fig. 10.3 Schematic editor and waveform viewer

10.3.2 Inverter Layout and LVS/DRC

This may not apply to those fortunate ones who can request layouts to layout experts, but if you are drawing your own layout, a simple layout should be done before completing the details of the circuit.

First, obtain a design rule manual with detailed rules as shown in Fig. 3.12, and skim over it. The first thing to check is the minimum grid size (Fig. 3.13).

After appropriately applying the necessary setting files for the layout editor, you should be able to bring up the layout editor (practically speaking, these kinds of steps can actually become bottlenecks). The minimum grid should be set in the layout editor (this can sometimes be written into the setting file, per layer).

By staring at the layout manual and fully utilizing the ruler functionality of the layout editor, an inverter layout should be drawn with the circuit diagram and the hierarchical structure shown in Fig. 3.25 in mind. Be sure to use the layers defined in the layout editor such as active area or METAL1. Here, as shown in Fig. 3.17, make sure to confirm with the automatically generated layers and complete the layout only with the minimum required layers. Also, as shown in Fig. 3.24, it is strongly recommended that the interconnect is split into different layers for signal, supply, and ground. Even if this distinction is not made in the given setting files, it is worthwhile to split the layers by searching and editing the appropriate place in the setting file. Although this can seem laborious, this is much simpler than taking the time and effort to go through layout and debug without doing so.

Once layout is finished, the GDSII file can be extracted, and LVS and DRC can be run. In a normal layout editor, the LVS or DRC error output files can be read in, and the corresponding locations can be highlighted as in Fig. 10.4. Make sure to tweak settings until this can be done properly. Also, the LVS options in Fig. 3.33 should be chosen at this stage.

Once LVS and DRC are confirmed, you should be able to complete the layout for a five-stage inverter, with the hierarchical structure in mind, and LVS/DRC should be run as well.

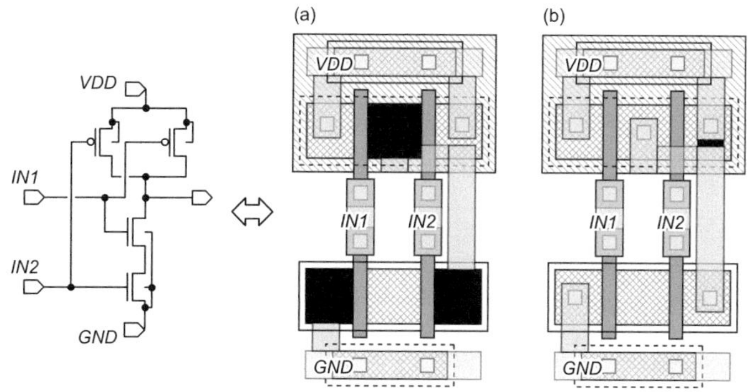

Fig. 10.4 Cross-highlighting: (**a**) LVS error, (**b**) DRC error

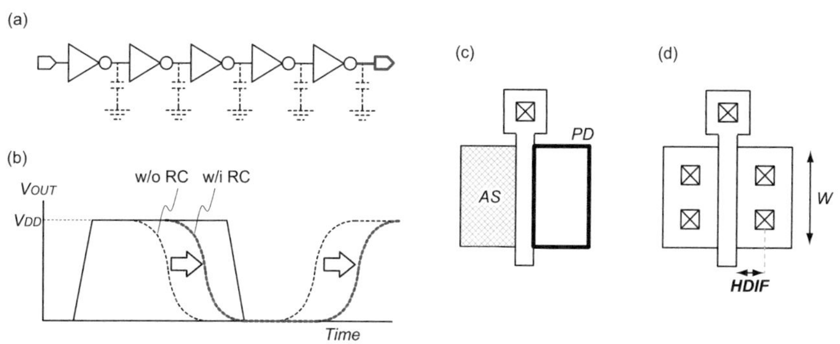

Fig. 10.5 Adjusting AD/AS/PD/PS

10.3.3 RC Extraction

RC extraction should be done on the five-stage inverter that you have just laid out. For circuit sizes of this order, there is not a significant effect on the simulation time even if the element count goes up. Therefore, both R and C should be extracted, as opposed to a simple C extraction. A SPICE simulation should be run with this netlist after RC extraction. You should be able to reuse the same simulation control file from a previous section. Here, the delay in the simulation results of the RC extracted netlist should be larger than the delay from the netlist without RC extraction, as shown in Fig. 10.5a, b. If the circuit design is completed with the netlist without RC extraction and then the layout is done, there is a high probability of disaster that the RC extracted netlist indicates a delay that is too large, and the circuit must be redesigned. To prevent this, some amount of the effects due to interconnect RC should be considered in circuit design. For this, it is recommended to imitate the effects of interconnect RC in advance by placing extra source and drain capacitance at the circuit design stage. That is, for the BSIM 3 model, the SPICE parameter file should be edited to increase the **HDIF** value, and as shown in Fig. 10.5b, this value should be adjusted so as to match the simulation waveforms of the netlists with and without RC (the simulated waveform for the RC extracted netlist serves as the baseline). Also, when using a model without **HDIF** as with BSIM4, and equivalent **HDIF**-like parameter should be internally defined, and by setting $AD = AS = 2 \times \mathbf{HDIF} \times W$, $PD = PS = 2 \times W + 2 \times \mathbf{HDIF}$ and by writing a script that adds these AD/AS/PD/PS values to the netlist, the internally defined **HDIF** parameter value should be adjusted to match the simulation waveform of the netlist with RC.

The objectives at this stage are to be able to properly extract RC from layout and to find the appropriate **HDIF** value to use at the circuit design stage. From here on, the value of **HDIF** found will be used to conduct the actual circuit design.

10.4 Finally, Some Real Design

10.4.1 Circuit Design and Considering the Measurement Methodology

The internal circuit can be designed with specific needs in mind. Here, we consider the exchange of signals with the world outside of the chip.

10.4.1.1 Supply and Ground

- How many types are needed? For analog, digital, and IO. Any further divisions?
- Will deep N-well be used? Will the grounds be shared? Will the wiring be separated even if they are connected in the substrate?
- How many pins each are needed? Is the current density not excessive? Are the parasitic inductances and capacitances sufficiently small?
- If working in teams, is there consistency across designs? What about the label names for LVS?

10.4.1.2 Input and Output Signals

- Which signals will be input from outside? What are the frequencies? Which device will they come from? What will we do about the termination (50 Ω/1 MΩ)? Are they analog or digital? What will we do about the IO buffers?
- Which signals will be output from the chip? What are the frequencies? Which device will they be measured with? What will we do about the termination (50 Ω/1 MΩ)? Are they analog or digital? What will we do about the IO buffers?
- Which test signals will be extracted to observe the internal circuit in case there is unexpected behavior? What about the noise effects due to extracting these signals?

10.4.1.3 Packaging or Probing

- What is the maximum signal frequency? Does a package suffice to input and output these signal frequencies?
- Will we probe the chip?
- Are there enough pins?

10.4.1.4 Internal Circuitry for Observation Purposes

- Is the equipment we have at hand enough to make measurements? (Are the equipments' observable frequencies high enough? Are there enough input and output ports on the equipment?)
- Does the package have good enough frequency characteristics? Are there enough input and output pins on the probe?

If these conditions are not met, circuits exclusively for measurement purposes must be included, such as a VCO or PLL to generate a high CLK frequency, or extracting a waveform by sampling and down-converting.

10.4.2 Layout Design

10.4.2.1 Determining Pin Placement

Wafer probing or packaging? If we decide to go with probing, the pad placement is determined by matching to the probe pin count and pitch. With high-speed signal probing, the pins should be in the order of GSG or GSGSG. With low-speed signal probing, sometimes the supply and ground pin locations are predetermined, and the pad placement must follow this. Also, probes can come from either two sides or four sides during measurement. The pin placement should be such that the probes do not interfere with one another but, at the same time, take up as little area as possible. Probing from four directions is more difficult during measurement.

If we decide to go with packaging instead, the pin placement should be determined with the precision of the processing and assembly of the board for the package in mind. First, we should go find something that looks like Fig. 10.6a and confirm the correspondence between the internal pads of the chip and the package pins. Based on this, for example, with Fig. 9.9, the two adjacent positions of a package signal pin are determined to be ground in a GSG configuration to make the signal routing easier on the board. If there are extra pins available, a GGSGG configuration will make soldering much easier.

Pin placement should be determined by considering what is shown in Fig. 5.16 as countermeasures against noise.

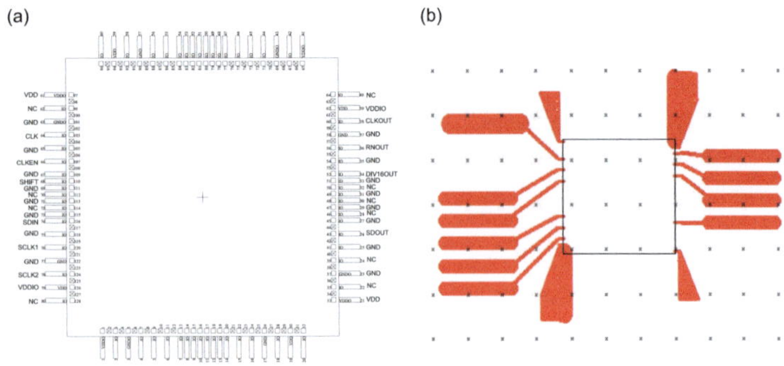

Fig. 10.6 Board design

10.4.2.2 Overall Layout of Supply and Ground

As shown in Fig. 5.15, rings are formed for both supply and ground for IO power. If there is a voltage conversion within the IO and, for example, the signaling with the outside is at 3.3 V but the internal circuitry is at 1.0 V, then we need to make a total of four rings, one for both supply and ground at both 3.3 and 1.0 V.

To avoid IR drop of the supply due to the resistive component of the internal circuitry, the wiring should be thick and in a mesh structure, as shown in Fig. 6.7c. Also, as mitigation against supply noise, decoupling capacitance should be placed if there is extra space. The optimal composition should be thought up while considering the leakage current and capacitance values, as well as the options and conformity with LVS.

From the perspective of separating noise, there should be a separate supply and ground for the IO that burns a large current, the large-scale digital circuit, and the low-noise analog circuit. If possible, a deep N-well structure as well as a guard ring as shown in Fig. 6.10 should be utilized.

10.4.2.3 Overall Layout Including Interconnect

By collecting all of the analog blocks and high-speed blocks into one location, it becomes easier to take care of special considerations.

Also, the wiring from IO to the internal circuitry tends to be extremely long, so the analog blocks and high-speed blocks should be placed near the IO to suppress effects such as wiring delay. Also, antenna rule violations tend to occur with these extremely long wires, and in these cases, some measures such as those indicated in Fig. 3.16 are necessary. It is quite depressing to see antenna errors toward the end of your design, and the deadline is staring at you in the face, but these must be taken care of. You should submit the design after all checks, such as LVS, DRC, ERC, antenna, density, etc., have passed.

Good work! Let's go have a drink.

10.5 After Submission of Design Data

I'm sure you're relieved that the design is over, but "you are a new person every day." While your memory is still fresh, let's finish whatever else we have to do.

10.5.1 Preparation for Measurement

Yes, I completely understand when you say "I don't want to get ready for measurements until the chip comes back." I understand, but it's more efficient if you complete these tasks now while your adrenaline is still rushing.

10.5.1.1 Creating the Board and Jigs

The steps for board design are indicated in Fig. 9.9. On Fig. 10.6a, which describes the connections between the internal and external pins, we write in the corresponding package pin and the signal name that is used in the internal circuitry. During measurements, this pin placement map, along with the circuit schematic, will always be at hand.

In addition, a board pattern such as that shown in Fig. 10.6b is designed based on the package form data, and special equipment is used to create the board. If you are going to ask an outside supplier to do this for you, then you will create the design data that the supplier requires and give that to them. Board design tends to be put off and delayed, and often times we ask "please do this ASAP" right before the deadline (maybe that's why some of the people from these board fabrication companies seem so strict). In any case, from sending them the design data to the delivery of the board will take at least one week.

Soldering cannot be done until the chip arrives, so here we complete tasks up to the data generation of Fig. 10.6b or until the board is fabricated. Once the chip has arrived, parts such as decoupling capacitors and various terminals and cables are soldered onto the board.

In measurements with probing, there is no need for board design, but jigs are required instead. In the header pin at the bottom right of Fig. 9.6d, lead lines for the power supply as well as decoupling electrolytic capacitors are soldered. It helps to make these also, while the pad placement and supply information is still fresh in your memory.

10.5.1.2 Creating Programs for Measurement

You should have been thinking about how to make measurements from the design phase, so now is the best time to start writing the basic programs for measurements.

It is necessary to search for the appropriate commands for operation from the Programmer's Guide while writing the program, especially if the equipment is being used for the first time. Thus, it can take an unexpected amount of time to control the equipment to your satisfaction.

However, depending on the measurement results, the chip can show unexpected behavior which may lead to additional measurement requirements. Considering that "you are a new person every day" in programming as well, perhaps this can be postponed until the chip arrives. However, it is still recommended to have at least some of the program written, especially in cases where there is not a lot of time from receiving the chip to the conference submission deadline.

10.5.2 Preparing Patent Documents

As opposed to the university setting, in industry, patent applications are necessary. One common pattern seems to be to write the patent after the design is finished but before the chip arrives. Although results cannot be presented at conferences unless the chip functions properly, patents seem to be filed regardless of whether the chip works or not.

10.6 Measurements and Onward

10.6.1 Measurements

Measurements are conducted once the chip arrives. This is a bland procedure that requires persistence. Thus, measurement results are precious, and conversing with an actual chip will yield new discoveries. Also, you should always remember that measurement results are "once in a lifetime" and make sure to fully utilize measurement automation programs to run through the measurement process.

10.6.2 Writing a Report

Measurement results should be summarized in an easy-to-understand manner by distinguishing between the "truth," which includes measurement methods, connection information, measurement conditions, and raw data, and the "observations" or conclusions that are drawn from the data.

If good results come out, they are submitted to a conference. The title, abstract, introduction, and conclusion should be written first for the manuscript. This is only about 1/5 of the total, but you could consider yourself to be halfway through.

In other words, these are the important sections and thus should be written with utmost care. The rest is relatively easy to write. When the acceptance e-mail with the "Congratulations!" message is received, you should begin the preparation of your presentation slides. Make sure these are simple and beautiful, and at the same time, the script for your speech should be written as well. Make sure to get enough practice and memorize the script so that you do not have to look at it during your presentation.

The next step is a journal submission, which should take into consideration any questions raised at the conference and also describe in detail the theoretical background. In most cases, conditional acceptance will come first before the final acceptance of your paper.

10.6.3 Toward Your Next Design

Once you are here, you probably thoroughly know about the advantages and disadvantages of the fabricated circuit more than anyone else (professors, bosses, etc.). You might come up with circuits that improve on these disadvantages or even a new circuit that solves the problem. Utilize your imagination and creativity to design better and better circuits.

Epilogue

With more freedom, more richness, and more happiness

Thank you very much for reading this text all the way through. I believe that by reading it, you have gained knowledge you did not have before and you can now understand the meaning of some important jargon. As a result, you should be more confident in your circuit design, and by relying not on your seniors, boss, or professor, but yourself, you have gained more freedom.

For us who live in a country with a free economy, a market of billions of dollars expands before our eyes, and with a single click we can access the market of the entire world. Also, relative to other industries, the semiconductor industry has a market that is the most open and the most fair. In the process of choosing our careers, we were very lucky to have chosen the field of semiconductor circuit design among the many technical fields out there. By acquiring the proper techniques, we can become freer, richer, and happier in this open and fair market.

If the contents of this text contribute to your knowledge and confidence and help to make you a free, rich, and happy circuit designer, I could not be happier as an author. More than anything, this makes it worthwhile for you to have gone through this text.

Toru Nakura
June 2011

T. Nakura, *Essential Knowledge for Transistor-Level LSI Circuit Design*,
DOI 10.1007/978-981-10-0424-7

MIX
Papier aus verantwortungsvollen Quellen
Paper from responsible sources
FSC® C105338

If you have any concerns about our products, you can contact us on
ProductSafety@springernature.com

In case Publisher is established outside the EU, the EU authorized representative is:
Springer Nature Customer Service Center GmbH
Europaplatz 3, 69115 Heidelberg, Germany

Printed by Libri Plureos GmbH
in Hamburg, Germany